Viruses

VIRUSES

Arnold J. Levine

**SCIENTIFIC
AMERICAN
LIBRARY**

A Division of HPHLP
New York

Library of Congress Cataloging-in-Publication Data

Levine, Arnold J. (Arnold Jay), 1939–
    Viruses/by Arnold J. Levine.
      p. cm.
    Includes index.
    ISBN 0-7167-5031-7
    1. Viruses. I. Title.
    [DNLM: 1. Viruses. QW 160 L665v]
  QR360.L38 1991
  616'.0194—dc20
  DNLM/DLC   91-20456
  for Library of Congress   CIP

ISSN 1040-3213

Printed in the United States of America.

Scientific American Library
A Division of HPHLP
New York

Distributed by W. H. Freeman and Company.
41 Madison Avenue, New York, New York 10010 and
20 Beaumont Street, Oxford OXI 2NQ, England

2 3 4 5 6 7 8 9 0 KP 9 9 8 7 6 5 4 3 2

This book is number 37 of a series.

To Linda, Samantha, and Alison

Contents

Preface

The last thirty years have witnessed an extraordinary revolution in biological science, inspired by the tools and techniques of molecular biology. The viruses and their hosts—in particular, bacteria and human beings—have been among the organisms most frequently and thoroughly studied by molecular biologists. These researchers have joined forces during recent decades with geneticists and biochemists—in a cooperative effort without precedent—to provide a detailed description of the structure, replication, and genetic information of viruses.

Had the retrovirus that causes AIDS (acquired immune deficiency syndrome) been discovered in 1961 instead of 1983, we could never have understood the nature of this disease in only three to four years. At that time we had identified neither the way in which the virus that causes AIDS duplicates itself nor the cell in the human body that it replicates within and destroys. We simply did not have enough information then to begin to explore AIDS at almost any level. By 1983 we did, because of the latest acceleration in the progress of our knowledge since the discovery of the first virus by Dimitrii Ivanovsky in Russia one hundred years ago (1892). Since that time, thousands of different viruses have been identified and studied, and a number of major human, animal, and plant diseases have been shown to be of viral origin. It is a good time to review that progress.

This volume has two goals. The first is to give an account of what we know about the life of viruses—a form of life that cannot reproduce outside a host cell and that has, therefore, evolved complex and varied interactions with higher organisms. Because there are so many different viruses, choices had to be made. The viruses of human beings were selected as the focus for all the later chapters, where particular virus groups are singled out for discussion and scrutiny. The distinct emphasis, moreover, upon viruses involved in causing or contributing to cancer reflects the interest, ideas, and knowledge of the author.

The book's second goal is to recount—for given crucial instances in the history of this field— how virologists learned what they did. I hope to convey that the real excitement of science is in its practice: testing the concepts, models, and hypotheses that drive the work. This book attempts to tell the one-hundred-year story of scientists' quest to understand the smallest and simplest forms of life. It has not been possible, or indeed desirable, to separate the viruses from the virologists—from the way in which scientists get ideas, test their validity, apply them or try again. The best result provides the next hypothesis for testing, in an endless path that is a pure joy to the scientist and has considerable utility to the human population.

No book is produced solely by one person, and this book received a great deal of support all along the way. Kate James helped to type and retype, reading every version and providing helpful and cheerful comments. Alice Lustig and Steve Holtzman read and commented upon every chapter, and both have had a significant impact on the final product. My scientific friends and colleagues who read selected chapters and provided critical input are Drs. John M. Coffin, Clyde S. Crumpacker II, Donald Ganem, Harold S. Ginsberg, Peter M. Howley, David Knipe, Robert A. Lamb, Peter Palese, Arnold B. Rabson, Bernard Roizman, and Robert Sinsheimer. Special thanks as well to all those researchers around the world who contributed photographs to this book; they are individually acknowledged in the illustration credits. I would be remiss, however, if I did not single out Dr. Babu Venkataraghavan for his heroic computer graphics and Dr. Frederick A. Murphy for his special support and efforts. The manuscript and its organization received a great deal of skilled attention from Amy Edith Johnson, the development editor; Diana Siemens, the project editor; and Travis Amos, the photo editor, all at Scientific American Library.

The time needed to carry out research and to write this book was made possible by a generous sabbatical leave from my duties at Princeton University. During this time Dr. Thomas Shenk, a good friend and colleague, did the hard work of chairing the Department of Molecular Biology at Princeton University. Throughout this sabbatical year, I was supported by grants from the American Cancer Society, the National Cancer Institute, The Guggenheim Foundation, and the ImClone Corporation. Finally, I wish to acknowledge my friend and host at the Pasteur Institute in Paris, Dr. Moshe Yaniv, who provided a chance to work with few interruptions.

Arnold J. Levine
Princeton, New Jersey, 1991

Viruses

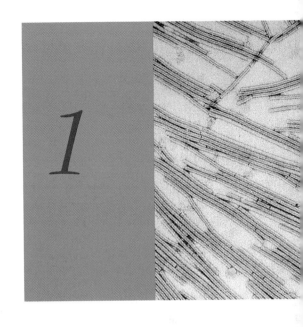

Viruses

Four tons of wheat; eight tons of rye; four fat oxen; eight fat pigs; twelve fat sheep; two hogsheads of wine; four barrels of beer; two barrels of butter; one thousand pounds of cheese; one bed, with accessories; one full-dress suit; and one silver goblet.

Goods traded in 1625 for one bulb of the "broken" tulips that became an aesthetic and commercial craze: the "tulipomania" of the mid-1630s in Holland

The human mind has always been fascinated with the concepts of largest and smallest. For mathematicians, these are infinity and zero; for physicists, the ever-expanding universe and subatomic particles. Biologists have a special feeling for whales and redwood trees, each of which has a grandeur all its own. At the other pole are the smallest of all living things, the viruses. Viruses populate the world between the living and the nonliving, the molecules that can duplicate themselves and the ones that cannot. Inherent in the organization and properties of viruses are many of the secrets of life and life processes.

All life forms employ two categories of chemicals, one that stores information and a second that acts, based upon that information, to duplicate the organism. The information stored in a living entity provides a plan or blueprint to carry out the organism's life functions and to pass a nearly exact duplicate of the plan to the next generation. Just as on a computer tape, the information stored in

Left: "Color breaking" in tulips, which created variegations with a random, paint-splashed character, was first described in western Europe in 1576. This detail, from Jan van Huysum's *Vase of Flowers,* is one of many Dutch paintings depicting the oddly beautiful outcome of infection with the tulip mosaic virus.

Above: The tobacco mosaic virus (electron micrograph), which also mottles its host, was the first submicroscopic pathogen to be identified, some three centuries later.

1

living organisms can be read linearly; at each position on the "tape," a bit of information is given. ("Bit," an abbreviation of the compound "binary digit," refers to the smallest unit of information stored on a computer tape or disk; here it corresponds to the nucleotide, the smallest unit of information in a chromosome.)

Consider the smallest of viruses, the viroids, which duplicate themselves in plants—and, in so doing, cause serious plant diseases. The simplest viroid contains only 240 bits of information, about ten million times less than the human information base (three billion bits). These 240 bits are arranged on a circular chromosome (the equivalent of the computer tape) and contain a set of signals that permit the molecule to duplicate itself. The replication of this viroid must occur within a plant cell, because the host cell contributes all the components needed for the viroid chromosome to make copies of itself. Indeed, all viruses can duplicate themselves only inside cells, because they require significant contributions from their hosts. But because different viruses take different things from their hosts, there is a great diversity in both viruses and the toll they exact—that is, the diseases they cause.

The diseases caused by viruses have had an enormous impact upon human beings. The existence of these parasites has shaped the evolution of plant and animal hosts and played a significant role in who we are today. Viral diseases, moreover, have been important events in our history. For example, it is unlikely that a small band of Spanish soldiers in Mexico could have defeated the Amerindians in 1520 in the absence of a raging smallpox epidemic brought inadvertently to the New World by the soldiers themselves. On the other hand, there has not been a single case (barring lab accidents) of smallpox in the world since October 1977; there is every reason to believe that an intensive vaccination program has now eliminated a disease that had a two-thousand-year recorded history. At the opposite extreme of our understanding and control, the present-day epidemic of AIDS, which is caused by the human immunodeficiency

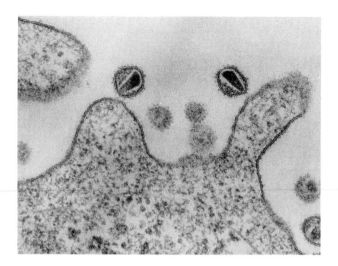

Electron micrograph (EM) of HIV, which causes AIDS, attaching to the surface of a T-cell lymphocyte. Magnified about 120,000 times. (EM magnifications throughout this book are between approximately 50,000 and 150,000 times.)

virus (HIV), is an extraordinary puzzle that will be with us for many years to come. Clearly, the medical consequences of virus infections remain an important reason to study these organisms.

During the last half of the twentieth century, a revolution in the biological sciences has taken place. New approaches and technologies have permitted us to study life processes at the molecular level. We have learned both to count and to determine the sequence of the bits of information (nucleotides) that make up the blueprint of life. The genetic code required to translate those bits of information into the molecules that act has now been deciphered. The development of molecular biology was led in a significant way by the study of viruses and their hosts. We have learned that many life forms and processes display common features. Because viruses depend upon their hosts to supply the tools needed to replicate themselves, the virus must share the same rules and signals with the host. Lessons learned from the viruses are thus applicable to more complex organisms.

Some viruses cause cancer in their hosts, by distorting the functions and signals used by the host cell to maintain itself. Identifying these functions and signals has been critical to our present understanding of the molecular basis of cancer, and the study of viruses has pointed the way to new approaches for treatment and prevention. An experiment is under way in Taiwan, for example, in which 63,500 newborn infants were recently immunized to prevent hepatitis B virus infection; the prediction is that forty or fifty years from now there will be 8300 fewer cases of liver cancer in that population. What we have learned from viruses has contributed to our understanding of life at both the molecular level and the population level: from the smallest perspective to the largest.

The first viruses were recognized as special entities only about a hundred years ago, between 1886 and 1898. We have come a long way since then, proceeding by asking one question at a time and building upon the answers. What are viruses? Are they alive? What do they look like? Are there many different kinds of viruses? Does each virus cause one specific disease? How do viruses cause diseases? How do viruses duplicate themselves? What have we learned about viruses that can be applied to humans? Each of these questions was asked for the first time in the context of a set of observations and a scientific perspective. In this book, we will attempt to preserve that context so that we can better appreciate the often remarkable insights of the successive questions and the often unexpected nature of the answers. In that way, we can recapture the one-hundred-year-old science of virology.

Historical Perspectives and Definitions

The idea of a submicroscopic world of the viruses could not have been grasped before the discovery of the microscopic world. Living organisms too small to see with the unaided human eye were first observed by Antony van Leeuwenhoek (1632–1723). Leeuwenhoek lived in Delft, Holland, where he held a political sinecure as custodian of the town hall. He was also a cloth merchant who used simple one-lens microscopes to examine draperies and cloth textures. Despite comparative isolation and a lack of formal training, Leeuwenhoek devoted a great deal of time and effort to his hobby of lens grinding, producing the best lenses then available: they magnified an object about 300 times. Leeuwenhoek turned his attention and interest to common objects and began to examine "rain-well-sea-and-snow water." In a series of communications to the Royal Society of London, Leeuwenhoek described his microbial world of "wee animalcules," including what we know today as the bacteria (spheres, rods, spiral shapes), protozoa, algae, and yeasts, as well as sperm, eggs, red blood cells, and

Antony van Leeuwenhoek.

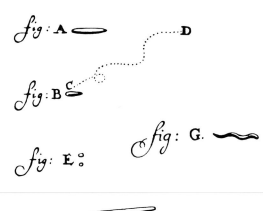

Leeuwenhoek's drawings of "wee animalcules."

With the invention and improvement of the two-lens, or compound, microscope, which had the ability to magnify one to two thousand times, six hundred distinct microbes were recognized in Ehrenberg's *Atlas* by 1838. (Today, the list of bacteria in Bergey's *Manual of Systematic Bacteriology* literally fills several volumes.) Once the microbial world had gained a firm reality, the scientists of the nineteenth century began to debate two questions whose answers would bring us closer to recognizing the existence of the submicroscopic world of viruses.

The first question—whether microbes can appear spontaneously in decomposing matter (that is, by "spontaneous generation") or arise from the duplication of similar microorganisms—was hotly debated, with experiments apparently supporting both sides of the controversy. Although many good experimental proofs eliminating spontaneous generation as a viable theory predate Louis Pasteur (1822–1895), he is usually given credit for show-

more. Leeuwenhoek's zest for discovery and his wonder at seeing the microbial world unfold are best conveyed by one of his letters to the Royal Society, about a decaying tooth he examined: "I took this stuff out of the hollows in the roots, and mixed it with clean rain water and set it before the magnifying-glass. . . . I must confess that the whole stuff seemed to me to be alive. But notwithstanding, the number of these animalcules was so extraordinarily great that 'twould take a thousand million of some of 'em to make up the bulk of a coarse sand-grain."

The 300-fold magnification employed by Leeuwenhoek is theoretically below the limit required to visualize bacteria; it is likely that he had to rely upon indirect light (now called darkfield illumination) to visualize even the outlines of these organisms, although he never stated that in any of his letters. Indeed, Leeuwenhoek was quite secretive about his tools and methods; only the best lens grinders, working independently, could repeat his observations. Because of these limitations, Carolus Linnaeus was able to recognize only six species of microbes as distinct entities in the group of animals that he classified under the name *Chaos* in 1767.

Louis Pasteur.

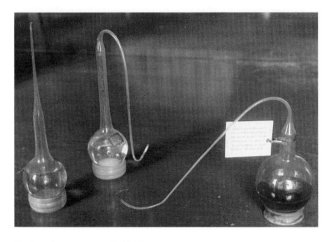

Pasteur's swan-neck flasks.

ing that boiled (sterilized) medium would indeed remain clear and free of bacterial growth in a specially designed "swan-neck" flask that allowed air access to the medium but did not permit airborne bacteria to enter. These vessels, with their original contents, remain free of bacteria to this day in the Pasteur Museum in Paris. Pasteur declared, in his usual confident manner, "Never will the doctrine of spontaneous generation recover from the mortal blow of this simple experiment"—and he was right. These new beliefs led directly to antiseptic surgery (through the efforts of Joseph Lister in the 1860s) and sterile techniques in medicine and in scientific research; isolating a single class of organism can be accomplished only by having a sterile field or culture within which to grow it.

The second great question tackled by nineteenth-century microbiologists was whether microbes cause specific diseases. From the time of Hippocrates (fifth century B.C.), miasmas, or poisonous vapors, had been invoked to explain epidemics of contagious disease. To distinguish a miasma from a microbe and to determine which one really causes disease were formidable tasks. The challenge was taken up by Robert Koch (1834–1910), who realized that what was needed was a good definition of "causative agent." Koch's postulates for distinguishing between a real pathogen (disease-causing agent) and a contaminant or adventitious microbe are employed to this day. These postulates state: (1) the organism must be regularly found in the lesions of the disease, (2) the organism must be isolated in pure culture (hence the need for sterile technique), (3) inoculation of such a culture of pure organisms into the host should initiate the disease, and (4) the organism must be recovered once again from the lesions of this host. In 1876 Koch proved by these means that anthrax in cattle was caused by the bacterium *Bacillus anthracis*. Koch went on to discover that these bacteria can form spores that are resistant even to boiling in water. This observation helped to explain why some experiments in spontaneous generation that had employed boiled water resulted in media filled with bacteria after the flasks had cooled; Pasteur's proof was thus reinforced. The existence of spores also explained why fields where anthrax-infected cattle had grazed could infect new herds even years

Robert Koch.

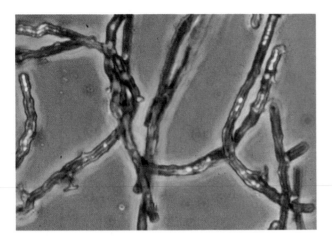

Bacillus anthracis.

later. The predictive value of these new sciences and the utility of their observations were becoming clear.

By the last decade of the nineteenth century, the microscopic world of bacteria was an important part of our understanding of disease, medicine, and agriculture, to say nothing of our newfound knowledge about bread, beer, and wine—the gifts of fermentation by yeast and other microbes. The microscope could resolve bacteria whose size was 1 to 2 microns (a micron is one millionth of a meter, or 0.0001 centimeter). The light microscope could, in the best cases, resolve a particle as small as 0.2 to 1.0 micron in diameter. The size of most common bacteria, 1 to 2 microns, was then the visual edge of the microscopic world, below which the human eye could not see. Armed with Koch's postulates, microbiologists began an intensive classification of bacteria and the diseases they cause. Their efforts went well until, in some cases, this already standard paradigm could not be repeated.

In 1886 Adolf Mayer, a German agricultural chemist working in Holland, was studying a disease of tobacco characterized by a pattern of light and dark areas on the infected leaves. He proposed the name "mosaic disease of tobacco" and proceeded to try to determine if this disease had an infectious origin. Mayer took the affected leaves and ground them up with water to produce a clear, soluble ex-

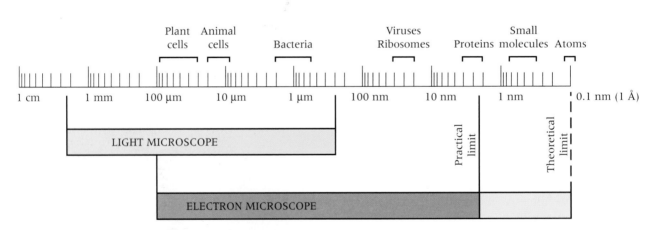

A logarithmic scale is used to compare the sizes of animal and plant cells, bacteria, viruses, proteins, molecules, and atoms. The light microscope can resolve objects of 0.2 micron. The electron microscope has a theoretical capacity of 1 to 2 angstroms, but its practical capacity is 2 to 3 nanometers. A micron or micrometer (μm) is 0.000001 (10^{-6}) meter. A nanometer (nm) is 0.000000001 (10^{-9}) meter. An angstrom is 0.0000000001 (10^{-10}) meter.

tract of leaf components, which he then injected into healthy plants. In nine cases out of ten, these plants showed all the symptoms of the disease. Mayer then attempted to culture, in pure form, the organism that caused the disease, using the standard techniques of his day for growing bacteria. Surprisingly, this search failed to isolate a bacterium or fungus in pure form, even though the disease could be transmitted like any other infectious disease. Koch's postulates could not be satisfied, and Mayer concluded that this disease was caused by a bacterium whose special nature prevented its culture, but which certainly would be revealed in future studies.

In the meantime, Chamberland filter-candles were produced—special filters that had pore sizes too small (about 0.1 to 0.5 micron) to let bacteria through. In 1892 a young Russian scientist, Dimitrii Ivanovsky, who was studying tobacco mosaic disease, reported to the St. Petersburg Academy of Science: ''I have found that the sap of leaves attacked by the mosaic disease retains its infectious qualities even after filtration through Chamberland filter-candles.'' The tobacco mosaic agent, which failed to replicate (that is, to increase in strength) in cell-free culture but appeared to grow on leaves, was smaller than anything previously described. (Ivanovsky pointed out the possibility that a toxin secreted by a bacterium caused this disease; toxins do not replicate themselves.) Six years later in the Netherlands, Martinus W. Beijerinck repeated Ivanovsky's experiment and showed that an infectious agent, able to replicate in the leaves but not in the filtered solutions, could transmit the tobacco mosaic disease. The submicroscopic filterable agent could be diluted manyfold and, when placed upon tobacco leaves, would produce many copies of itself that could be diluted again and shown to transmit disease. This plant pathogen, now called tobacco mosaic virus, was the first living, replicating member of the submicroscopic world to be recognized.

These observations were rapidly followed by the isolation of a filterable animal virus, the foot-and-mouth disease virus of cattle, by F. A. J. Löffler and P. Frosch in 1898. The first human disease

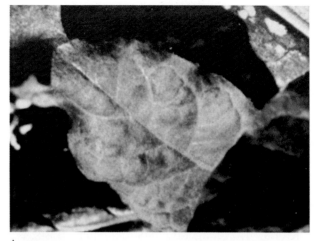

A

B

Tobacco leaves infected with the tobacco mosaic virus.
A: An early systemic infection six days after inoculation.
B: A late systemic infection four weeks after inoculation. Note the white spots, each of which is a site of virus infection, and the necrotic lesions.

shown to be caused by a filterable agent smaller than any known bacteria was yellow fever, recognized by the U.S. Army Commission under the direction of Walter Reed in 1900. Transmissible and filterable agents were even found in clear fluid extracts of bacteria themselves by Frederick W. Twort in England (1916) and Felix d'Hérelle in France (1917). D'Hérelle named these viruses of bacteria *bacteriophages.*

The term *virus,* which comes from the Latin word for poison, had been used synonymously for infectious agents of all kinds throughout the nineteenth and into the twentieth centuries. Beijerinck, in his discovery of the tobacco agent, called it a "contagium vivum fluidum." By the 1930s, scientists routinely used the term *filterable virus* to refer to any agent that passed through a filter fine enough to retain bacteria, but today *virus* is applied only to submicroscopic agents (less than 0.3 micron) that can pass through filters that retain most bacteria. These agents, whether their hosts are plants, animals, or bacteria, are obligate intracellular parasites—that is, they are able to duplicate themselves only inside a host cell; as the first experiments had shown, viruses cannot replicate in a cell-free (fluid) environment. This parasitism may result in the death or alteration of the host cells, which is in large part the reason that viruses cause diseases. For the early virologists, unable to see viruses in their light microscopes, there was an element of faith in these studies. But many viruses have now been shown to satisfy Koch's postulates: like the bacteria that cause anthrax, they can replicate and cause disease.

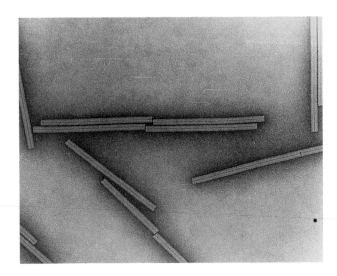

EM showing the rod-shaped virus particles of tobacco mosaic virus. The striations along the width of the rods reflect the helical symmetry of the subunits that make up the protein coat.

Modern Definition of Viruses

By the 1930s viruses were classified according to the host that they infected. Three groups of viruses—plant, animal, and bacterial (the bacteriophages)—were recognized; they were identified and stored in liquid filtrates derived from the leaves, lesions, or juices of their hosts. As methods became available to purify viruses from the complex mixtures of chemicals found in these fluids, the chemical composition of a virus could be determined and used to classify it.

When, in 1935, Wendell Stanley succeeded in obtaining a crystalline form of the tobacco mosaic

virus, this form was seen by some scientists as filling the gap between the living world of reproducing viruses and the chemical or nonliving world of molecules and crystals. In truth, the ability to form a crystal simply reflects the structural uniformity or symmetry of all the viruses in a population, together with a surface complementarity that permits an orderly aggregation of virus particles. German technology in the 1930s provided a new tool: the electron microscope. Instead of visible light, it employs beams of electrons focused by magnets—permitting the useful resolving power to go from one to two thousand diameters magnification to three hundred thousand diameters. Now the submicroscopic world of viruses could be visualized, and electron micrographs showed the beautiful symmetry of viruses predicted by the crystallization of the tobacco mosaic virus. Today, chemical composition, symmetry, and structure make up the basis of virus classification.

Almost all viruses are composed of two parts: a nucleic-acid core containing the genetic information and, surrounding and protecting this core, a coat built from repeating protein subunits and, in some cases, further encased in a lipid (fatty) envelope. A virus's genetic information may be encoded in deoxyribonucleic acid (DNA), as it is for all non-viral living organisms, or in some viruses it may be encoded in ribonucleic acid (RNA). Both DNA and RNA are long, linear polymers composed of subunits (monomers) that are linked together by strong chemical bonds, as we will see in Chapter 2. The monomers, called nucleotides, are each composed of a base, a sugar, and a phosphate group. The type of sugar gives the nucleic acid its name: in

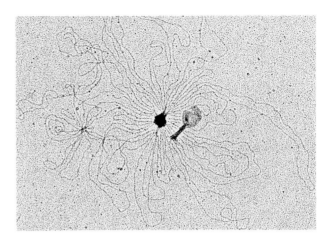

EM of bacteriophage T4; note the length of the DNA (released from the head by osmotic shock).

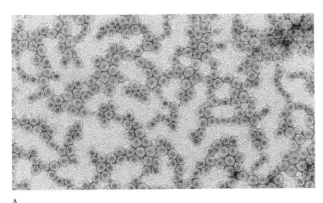

A

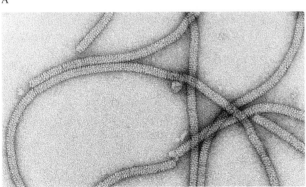

B

All plant viruses are RNA viruses. A: The isometric or spherical bromegrass mosaic virus, isolated from barley. B: The filamentous, helical beet yellowing virus, isolated from sugar beet.

DNA the sugar is deoxyribose, and in RNA it is ribose.

As we will examine in some detail in Chapter 2, DNA is usually found not as a single polymeric strand, but as a double strand with the two polymers wound about each other in a helix. The sequence of nucleotides in a DNA or RNA polymer determines the genetic information, and the chemical composition of this genetic information—whether it is encoded in DNA or RNA—is used to classify viruses. Viruses may contain double-stranded DNA (herpesviruses), single-stranded DNA (parvoviruses), double-stranded RNA (reoviruses), or single-stranded RNA (poliovirus). The RNA viruses are unique in that they are the only living organisms that use RNA to store their genetic information; all the other reproducing forms of life employ DNA. (The consequences of this will become clear when we discuss the RNA viruses in greater detail.)

In the simplest cases, the DNA or RNA of a virus is surrounded by proteins. Proteins are also polymers, composed of twenty different monomeric subunits: amino acids. The sequence of these amino acids determines the chemical properties of the protein by guiding the polymer to fold into a shape or structure that determines the protein's function. The sequence of nucleotides in DNA or RNA polymers specifies, in turn, the sequence of amino acids in a protein. In this way, genetic information determines a protein's sequence of amino acids, shape, and activity (function).

The function of a virus's coat protein is to form a cage or protective sphere around the genetic information (DNA or RNA) so that it is not attacked or altered by the environment. The protein coat also mediates the spread of a virus from host to host, ensuring that new viral generations will perpetuate themselves. Because a single protein is simply not large enough to encompass all the genetic information, hundreds or thousands of identical protein subunits are produced in the virus-infected cell. Several of these proteins come together into a symmetrical assembly unit called a capsomere. Capsomeres interact with each other through contacts on the protein surfaces to form a shell or coat surrounding the viral nucleic acid. This shell is often called a capsid, and the entire particle is called a nucleocapsid (DNA plus protein or RNA plus protein). The simplest viruses contain only this nucleic acid surrounded by a protein coat. More complex viruses surround the nucleocapsid with a lipid envelope and insert additional viral proteins into the envelope. A completed or mature viral particle is called a virion.

Different viruses encode the information in their DNA or RNA for distinct proteins that make up the capsomeres, capsids, and virions. Because of this, each virus has a unique set of capsomere proteins whose shape and composition determine how it will interact with other identical capsomeres and form a shell about the nucleic acid. Different viruses, therefore, produce virion particles with distinct shapes, sizes, and properties—each determined by the shape of the capsomere and its bonding patterns with neighboring capsomeres. Tobacco mosaic virus, for example, is composed of a single strand of RNA that produces a capsomere that assembles around the RNA to form a helical, rod-shaped structure. Other viruses, such as the rhinoviruses that cause the common cold, also con-

A

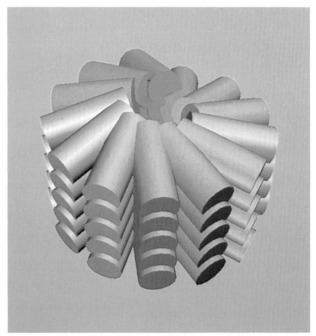

B

Computer-graphic representations of the tobacco mosaic virus capsomere. A: Each separate yellow line traces the amino-acid chain of a single protein. B: Capsomeres assembling into helical cylinders (virus particles).

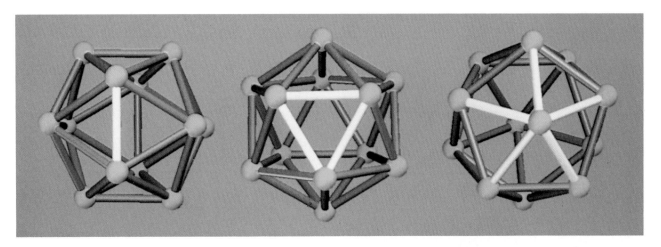

Computer-graphic representation of simple icosahedrons, looking down a twofold, threefold, and fivefold rotational axis.

tain a single strand of RNA, but the capsid is assembled into a spherical shape around the nucleic acid.

These closed-shell spherical virions have structures based on icosahedral symmetries. An icosahedron is an enclosed surface composed of twenty equilateral triangular faces. The symmetry of the icosahedron is characterized by rotational axes: each of the twelve vertices has a fivefold rotational axis (each of five rotations of 72 degrees produces an identical view), a threefold rotational symmetry around the center point of each of the twenty triangular surfaces, and a twofold rotational symmetry about each edge of the icosahedron, where two triangular surfaces meet. As recognized by the innovative architect Buckminster Fuller, who used the icosahedral shape to build his structurally sound geodesic domes, this class of polyhedron is the most efficient of all the closed shells that can be constructed. This is because the icosahedron uses the smallest subunits (capsomeres) to build an enclosed shell of a fixed size, which in turn conserves the genetic information needed to encode these subunits, resulting in a smaller DNA or RNA

Computer-graphic representation of a rhinovirus particle. Each colored sphere represents one subunit; these fit together to form an enclosed shell. The three pins are located at the points of fivefold, threefold, and twofold rotational symmetry.

strand. In addition, small protein subunits conserve the energy it takes to synthesize proteins in a cell. There are sixty identical elements on the surface of any icosahedron, and these elements form a symmetrical structure related by the twofold, threefold, and fivefold rotational symmetries.

Buckminster Fuller surrounded by models for icosahedral geodesic domes and enclosed spaces. Years later, virologists employing EMs of viruses built the same icosahedron models to reconstruct the morphology of viruses. Entering a geodesic dome would be like entering a virus particle.

Today, viruses are classified by their hosts, chemical composition (nucleic acid, protein, presence or absence of lipid envelope), shape, size, and symmetry. Such a classification is presented in the table in the appendix, which emphasizes the viruses discussed in this book. Included in this table are the diseases or associated pathologies caused by some of these viruses.

Cellular Structures and Functions

To understand viruses, we must first review our knowledge of the indispensable environment for viral replication—the living cell. Just as viruses are built largely from repeating identical subunits (capsomeres) that form symmetrical structures, so too the body of a multicellular host is built essentially from repeating units of cells. There are several advantages to building a structure from identical repetitive units. First, it is economical to mass-produce the same part over and over again. Second, if a rare mistake is made, only one of a million units is different; if it doesn't fit into the symmetrical lattice (because it *is* different), it usually doesn't hurt the structure. If a systematic mistake is made, on the other hand, that may be lethal. In addition, self-assembly systems are guided by a small number of attractive forces between identical subunits, repeated over and over again, to provide strength in numbers. Such systems build the symmetric structures so often seen in living organisms.

We see these principles in action at almost every level of life. At the molecular level, proteins assume their shape from their chemical composition; oppositely charged amino acids in different portions of a protein, for example, may cause it to fold, bringing together the opposite charges by physical attractive forces. The shape of a protein provides surfaces for aggregation with other proteins to form multimeric subunits in cells or capsomeres in viruses. With viruses, identical capsomeres then assemble to form a particle; virus particles may come together to form a visible crystal, as in the case of the tobacco mosaic virus.

Similarly, parts of a cell are built from repeating units that are organized to produce subcellular organelles where specific functions are carried out. A cell collects its organelles into a package by surrounding these units in a lipid membrane, the plasma membrane. Thousands and millions of identical cells are brought together to form an organ that represents the sum of these cellular functions. Identical repetitive cellular units in an organ are often interrupted by unique cellular subunits that carry out specialized functions. Organs or systems of functions are combined to produce a body, which gains its unique qualities by integrating these organ systems and having them work in a

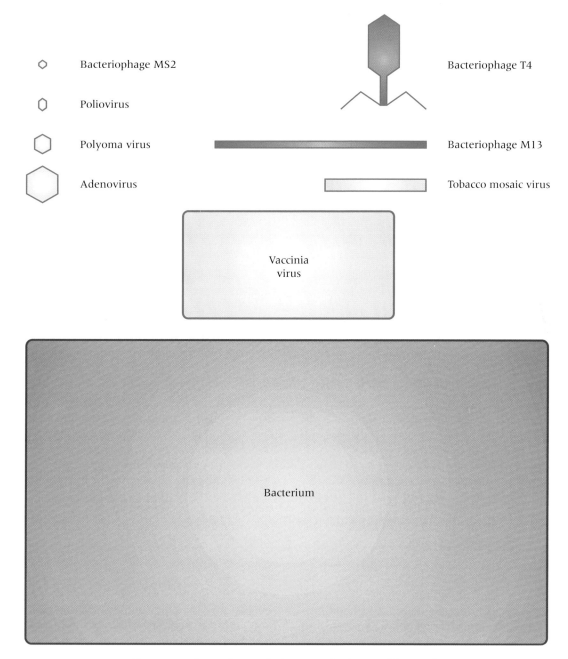

Bacteriophage MS2

Poliovirus

Polyoma virus

Adenovirus

Bacteriophage T4

Bacteriophage M13

Tobacco mosaic virus

Vaccinia
virus

Bacterium

Size comparison of several viruses and a bacterial host cell.

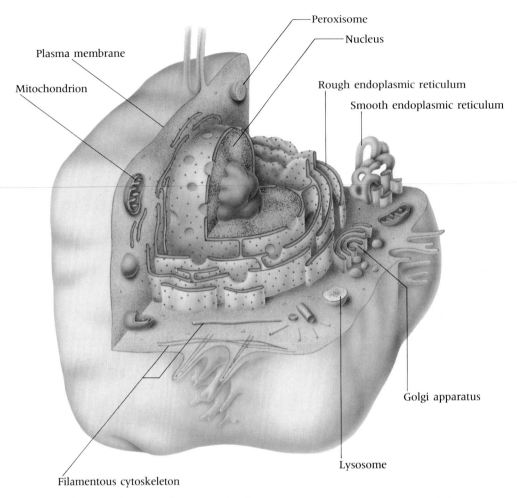

Peroxisome

Nucleus

Plasma membrane

Rough endoplasmic reticulum

Smooth endoplasmic reticulum

Mitochondrion

Golgi apparatus

Lysosome

Filamentous cytoskeleton

Schematic diagram of an animal cell; its components are a composite of the organelles found in several cell types.

coordinated way useful for the whole. It is clear that similar principles have evolved and been employed through the many levels of organization of life—from molecule to organelle to cell to organ to organism.

At the heart of any cell is the information needed to build the cell. All information is encoded in the DNA molecule housed in the cell nucleus. The first problem a cell or virus faces is that the DNA molecule is very long. In the case of a virus with 5200 nucleotides, the linear DNA polymer is several microns long, while the three billion nucleotides in human DNA are a meter long when stretched end to end. Written out as a linear series of one-letter abbreviations, the code or nucleotide sequence for a single virus would take a whole page of this text. The code for a human being would take five hundred thousand pages.

Packaging the genetic material, then, raises a problem. The diameter of a small virus may be

0.045 micron, and a cell nucleus with its DNA is only about 7 to 10 microns across. How can such a long molecule of DNA be fitted into such a small volume of space, while still leaving it available to replicate? The solution, evolved in both viruses and cells, is to wrap the long DNA strand around a protein core. This is not too different from wrapping a computer tape about a spool. For all cells, the protein core is called a nucleosome; it is composed of eight protein subunits (histones) that assemble into a nucleosome core. These nucleosomes then condense to form a linear filament, which in turn forms a long series of loops, back and forth, packaging the DNA into tight protein-DNA (nucleoprotein) complexes.

This DNA-protein package is separated from the rest of the cell by an envelope, the nuclear membrane. This is really a double-membrane system that is interrupted at periodic intervals by large pores where the outer and inner membranes are continuous. Both the nuclear membrane and the pores are highly selective, letting out of the nucleus some—but not all—molecules made there and taking into the nucleus some—but not all—molecules made in the cytoplasm (the cell protoplasm outside the nuclear membrane but bounded in turn by the cell's plasma membrane). Molecules—for example, some proteins—contain signals, determined by their sequence of amino acids, that govern entry into the nucleus. These signals tell each protein where to localize in a cell; they are the ZIP codes of a molecule. By this mechanism, proteins like the histones reside only in the nucleus, while other proteins reside only in the cytoplasm. This is a key part of the self-assembly system of a cell. Molecules must "know" where to go in a cell in order to build a structure that has spatially distinct elements.

The nucleus of a cell, then, is separated by its membrane from the cytoplasm, which contains a number of structurally and functionally distinct organelles. At several places, however, the outer nuclear membrane appears to be continuous with an extensive network of membranes in the cytoplasm, the endoplasmic reticulum (ER). This system of membrane-enclosed spaces seems to be used for the transport of certain molecules around the

EM of a thin section through a cell from the pituitary gland of a rat.

cell, allowing the placement of a molecule in a particular cellular locale or compartment. Lining the inside portions of the ER in the cytoplasm are ribosomes—organelles made up of subunits of RNA and protein (nucleoprotein complexes) that are the sites upon which all protein synthesis occurs. Newly made proteins on a ribosome can be transported across the membrane of the ER into the fluid-filled spaces of the ER, where specialized vesicles transport proteins to specific organelles and locations in the cell. (A vesicle is a spherical particle made of lipid, surrounding a fluid-filled space—a kind of bubble.)

Places where ribosomes collect on the ER are called the rough ER; they synthesize proteins for export outside the cell or for transport to specific cellular locales. Some of the cellular ER elements do not have associated ribosomes (the smooth ER), and some ribosomes are free of ER elements and synthesize proteins that remain in the cytoplasm. Once again, a spatial feature—where a protein is destined to reside in a cell—is associated with the place it is synthesized; that is, the rough ER exports the protein or localizes it to membranous portions of the cell, while free ribosomes synthesize cyto-plasmic proteins. Membranes thus serve to segregate the chemical reactions of a cell into compartments. The ER is also the site where new membranes are synthesized to increase the surface area on which chemical reactions may occur and to produce millions of vesicles to transport proteins about the cell for localization in the three-dimensional structure.

Some of these vesicles transport proteins to the Golgi apparatus, a system of membrane compartments organized into a stacked array. The different layers of membranes contain different sets of enzymes whose function is to add complex sugars to proteins or lipids, forming glycoproteins or glycolipids. This is often an important part of the preparation for either the secretion of a protein from a cell to the outside environment or the insertion of a protein into the plasma membrane of the cell. Modifying a protein by adding a sugar to it alters its chemical properties and may change its function or location in a cell. So-called exocytic vesicles (small, spherical lipid particles) bud from the Golgi and go to the cell surface. They fuse with the plasma membrane and secrete their contents to the environment outside the cell. In this way, cells communicate with each other, and the whole becomes greater than the sum of its parts.

Reciprocally, the plasma membrane is constantly budding new vesicles that contain elements of the outside environment *into* the cytoplasm; in this way, substances at the cell surface are transported into the cell and sampled to collect information about events occurring outside. After the infor-

mation is processed, the so-called endocytic vesicle has fulfilled its task. But the cytoplasm of a cell contains a number of vesicles with other specialized functions: lysosomes, for example, are membrane-enclosed bodies that store enzymes that digest or degrade proteins, DNA, or RNA. The lysosomal membrane keeps these powerful enzymes away from the cytoplasmic cellular proteins essential to life. Once an endocytic vesicle's sampling task is complete, it fuses with a specialized lysosome (an endosome), and the contents of the vesicle are degraded. The monomers formed as a product of this digestion (amino acids, nucleotides, and so forth) are then reused in the synthesis of new molecules.

Although lysosomes are only a small part of a cell, when they do not function properly the entire organism is at risk. Tay-Sachs disease is an inherited disorder in which the lysosomes that digest lipids are missing a single enzyme needed to break down these fats. The resulting accumulation of lipids blocks nerve-cell impulses, and the consequences are fatal to the organism.

Viruses frequently take advantage of the target-specific or vesicle-specific signals (ZIP codes) that direct molecules to the lysosomes. Semliki Forest virus, a togavirus that replicates in both insects and humans, has a coat protein that permits it to attach to a specific receptor at the plasma membrane of a cell. This virus is taken into the cell in an endocytic vesicle that, by virtue of the ZIP code on the viral protein, targets the vesicle for fusion with an endosome. The viral-encoded coat protein has stolen the same chemical signal used by cellular proteins to fuse the endocytic vesicle to the endosome. The fusion of these two vesicles releases the virus into the cytoplasm, where the viral RNA uses the host's ribosomes to synthesize viral proteins. Then the virus uses the host's ER, Golgi apparatus, and exocytic vesicles to exit from the cell via the plasma membrane. The virus has ensured its own propagation by using the same ZIP codes as the host proteins to exploit cellular organelles.

Two more dimensions of cell function demand brief attention: energy supply and three-dimen-

EM of the surface of Sindbis virus, a close relative of Semliki Forest virus. This sample was frozen and the lipid membranes cleaved in a technique called freeze-fracture, which permits a view of the virus surface projections and the inner membrane surface.

sional structure. The synthesis of proteins, the traffic of these vesicles and proteins about the cell, the duplication of the genetic information, and the constant synthesis of new vesicles are all processes that require enormous amounts of energy. The synthesis of polymers from their respective monomeric components is common to all four of the major chemicals in a cell: amino acids produce proteins, nucleotides produce DNA or RNA, sugars produce polysaccharides (complex sugars), and fatty acids produce lipids. In each case, synthesis consumes energy to link molecules together and generate polymers that store information or act biochemically. All cells, in all organisms, use the same currency of energy—a chemical that can channel its energy into heat (to keep body temperature constant), movement, synthesis of new polymers, and duplication of the cell and organism. (Even the processes

of reading and thinking about this consume chemical energy.) The chemical in the cell that does all this is called adenosine triphosphate (ATP). This is a nucleotide that is also used as a precursor of RNA. ATP stores energy in its chemical bonds, especially the bonds that hold the three phosphate groups together. When ATP is degraded to adenosine diphosphate (ADP), energy is released that can be transferred to other molecules that effect movement or thought, assemble new polymers, or produce heat.

Cells have perfected systems to synthesize tremendous levels of ATP—a single healthy liver cell will synthesize ten million molecules of ATP every second. Given the billions of cells in the human body, we produce and consume staggering amounts of this chemical fuel all the time. ATP is generated during a series of reactions linked to the complete breakdown of glucose (a sugar) into carbon dioxide (we exhale it) and water (we excrete it). In the best cases (that is, with lots of oxygen around), we gain about thirty-six ATP molecules for every molecule of glucose converted to carbon dioxide and water. Most of our ATP is produced in a specialized organelle in the cytoplasm called a mitochondrion. Each mitochondrion has a convoluted membrane system upon which resides a set of enzymes that function as pumps, gates, and channels to transfer the energy gained from the breakdown of glucose into ATP. ATP is then translocated out of the mitochondria into the cytoplasm for use in hundreds of reactions that fuel the cell.

This picture of the compartmentalization of a cell—employing membranes to separate functional units and using ZIP codes, in the form of chemical sequences, to transport proteins from organelle to organelle—reflects the high degree of internal cellular organization. Most cells can change shape in response to external stimuli, repositioning their internal organelles and, in some cases, moving from place to place. The ability to accomplish these activities depends on a network of protein filaments and tubules in the cytoplasm. These elements are termed the cytoskeleton of the cell. Not surprisingly, both filaments and tubules are built from protein subunits that can assemble and disassemble rapidly in a cell. This process is controlled at several levels; the building of cytoskeletal elements at one position in a cell while disassembly proceeds at another location is the basis for shape changes, cell movement, or the redistribution of organelles. ATP-to-ADP conversions release enough energy to slide one protein filament against another, generating a force for movement. While all cells have filaments and microtubule arrays, certain specialized examples of these subunit assembly systems in, for instance, muscle cells have been adapted for particular purposes.

The cell that we have been describing here is an average, nonspecialized cell from an animal or a human being. Plant cells have the same organelles, plus additional ones to collect sunlight and convert carbon dioxide and water into glucose. Bacteria, however, are quite different. One of the major distinctions between bacteria (and their relatives) and all other plants and animals is that the bacterial nucleus is not surrounded by a membrane—that is, the DNA lies in the cytoplasm without the separation provided by the nuclear membrane. In addition, bacteria do not have clearly identifiable sets of subcellular organelles like the ER, Golgi, or mitochondria. These differences are so fundamental that bacteria and their relatives are classified separately from the plant and animal kingdoms and are designated prokaryotes (having a primitive nucleus). Cells of animals or plants are designated eukaryotes (having a true nucleus). The advantages of specialization by the more highly evolved animals and plants do not result only from the differences between cells in a multicellular organism. As we have seen, the higher animal or plant cell is more specialized within itself, when compared to the cells of prokaryotic organisms.

Steps in the Replication of Viruses

Viruses are not cells, but they require cells. As obligate intracellular parasites, they must attach to and enter a host cell in order to reproduce. Although different groups of viruses have evolved diverse strategies to replicate themselves, it is possible to review the events in the life cycle (replication cycle) of a generalized virus to provide a background for understanding individual virus groups.

Virtually all viruses must accomplish eight key steps in their replicative cycle.

Step 1. The virus attaches itself to its host cell.

Step 2. The virus or its genetic information penetrates the cell.

Step 3. The nucleic acid is uncoated, which frees the DNA or RNA from its capsomeres or lipid envelope and permits the host cell to read out (express) the genetic functions of the virus.

Step 4. At this stage in the life cycle of many viruses, only a portion of the viral genetic information is expressed, resulting in the synthesis of only the subset of viral-encoded proteins collectively called the early viral gene functions (proteins). These proteins may function in one of several ways. In some cases, they contribute directly to the repli-

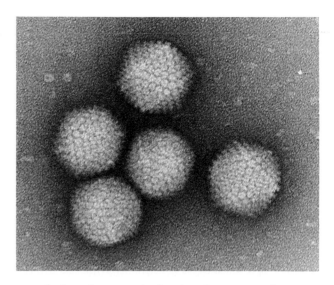

EM of adenovirus type 2, showing the capsomeric organization of the icosahedral virus particles. Small hexagonal objects in the field are loose capsomeres.

cation of the viral chromosome. In other cases, these viral proteins turn off many of the host-cell activities, maximizing the cell's available resources for virus production. Alternatively, some viruses that can duplicate themselves only in actively dividing host cells produce proteins that stimulate host-cell division.

Step 5. The viral nucleic acid is then synthesized to produce hundreds or thousands of copies of the viral chromosome.

Step 6. At this time, a second subset of the viral genetic information, commonly termed the late proteins, is expressed. These are the structural proteins, including the capsomeres of the virus.

Step 7. The capsomeres are assembled to form a shell around the nucleic acid of the virus.

Step 8. The mature virus, having duplicated its new copies, is released from the infected cell to attack a new cell and repeat this process.

Going from steps 1 through 8, a viral replicative cycle displays temporal organization: specific events occur in sequence, each dependent upon the successful completion of the previous step. This developmental process, which results in the synthesis of thousands of viral particles, is formally similar to the development of a multicellular organism from a single fertilized egg cell. Both processes require quantitative and qualitative changes over time. Early viral proteins, made from the single genome of the infecting virus, are expressed at low levels, while the late proteins are made from the newly replicated viral chromosomes produced in step 5. Quantitative changes are thus initiated and regulated chronologically.

Let us review each step in a little more detail.

Attachment or Adsorption

A virus uses specific proteins on its coat to recognize and attach to specific receptors on a cell surface (the plasma membrane). These receptors may be proteins or other components that are located only on certain cells; in effect, a virus may be able to attach only to a liver cell or to a lung cell and to

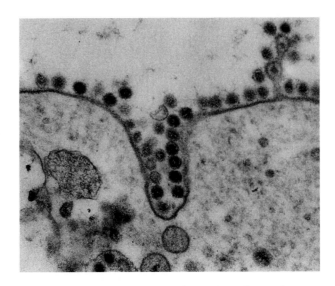

EM showing coronavirus particles (spherical) attaching to their receptors on the outer or plasma membrane (dark line) of a cell. The cell has begun to invaginate, forming an endocytic vesicle that engulfs the virus particles.

no other cell of the host body. This may result in specific disease states, such as hepatitis or pneumonia caused by viruses that replicate only in liver or lung cells. Viruses are said to have a specific tissue preference (tropism), and in some cases this is due to tissue-specific receptors.

The human immunodeficiency virus (HIV), for example, attaches to a receptor called the CD4 protein. The CD4 protein is found on the surface of certain lymphocytes (white blood cells) that are critical for the vitality of the immune system in humans. By entering into and killing only cells that have the CD4 protein on their surfaces, HIV kills the cells that maintain our ability to protect ourselves from infection, thus causing AIDS (acquired immune deficiency syndrome; see Chapter 7). The tropism of HIV is determined by its adsorption to cells with the CD4 receptor. Few other animals contain the CD4 protein on their cell surfaces (it is specific for some primates), so the HIV agent grows only in these animals (chimpanzees and humans).

Thus, the species limitations of some viruses may be due to restrictions in their ability to attach to specific cells. Other viruses replicate in many animals (influenza virus, for example), and this too can have profound consequences for the biology of the virus (see Chapter 8). There are also viruses that can replicate in many tissues or cell types of an animal, which means they use receptors that are found on most cells. Such viruses may cause widely disseminated disease throughout the body of the host.

Bacteriophages infect and kill bacteria, which are single-cell organisms. Like all viruses, bacteriophages attach to a specific receptor on the surface of their host cell. Occasionally, a mutation (a change in the genetic information) occurs in the bacterium so that the host cell can no longer synthesize the receptor on its surface. In many cases this is not detrimental to the bacterial host, which is then resistant to the virus that normally uses that receptor. As might be expected, resistant host cells arise in populations under attack by viruses—and these resistant cells, without receptors, survive, replicate, and eventually take over as a majority type in the population. In this case, resistance to virus infection is selected for by the presence of a killer virus.

But it is not in the best interest of the virus to kill all its host cells, leaving only resistant bacteria; if there are no hosts, there can be no viruses. Some viruses have developed alternative strategies for a live-and-let-live viral life cycle. Others have taken advantage of rare mutations in the virion coat protein, which may permit the virus to attach to a new receptor, even one found on the bacterial cell that was resistant to the original or parent virus. This rare virus is then selected for, by virtue of its newly acquired ability to attach itself to the otherwise resistant bacteria and duplicate itself in this new host. Host-range mutations, as they are called, that extend or restrict the ability of a virus to attach to susceptible host cells can thus occur in either the virus or the host cell. A virus evolves by altering its host range to be able to enter new environments. The attachment step provides a specificity and a selectivity that have profound consequences for the life cycle of a virus.

Penetration

For those animal viruses with lipid envelopes, penetration of the nucleocapsid core of a virion into a cell occurs when the virion envelope fuses with the plasma membrane to which the virus is attached. The fusion of viral and cellular lipid membranes is usually mediated by proteins encoded by the information (chromosome) found in the viral particle. Fusion leaves the nucleoprotein core of the virus on the inside of the cell; this penetration step is also part of the uncoating of such a virus.

Animal viruses without a lipid envelope (the so-called naked virions, composed only of protein plus DNA or RNA) are usually taken into cells to which the virus is attached by a process called phagocytosis, or endocytosis. As we saw earlier in this chapter, many cells constantly produce cytoplasmic vesicles by pinching off portions of their own membranes facing the outside of the cell to yield spherical vesicles that sample the extracellular environment. When these endocytic vesicles migrate to the cytoplasm and fuse with the endosomes, they transport a virus to a cellular location where the viral protein subunits are removed and the viral nucleic acid becomes accessible to the cellular environment. Here too, penetration and un-

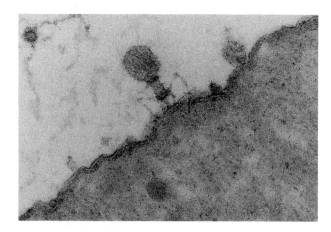

EM of a T4 bacteriophage adsorbed onto an *E. coli* bacterium during viral DNA injection into the host cell.

coating of the nucleic acid are coupled steps. In some cases, viral nucleocapsids are transported directly to the host-cell nucleus, where the viral chromosome resides during the entire life cycle.

Some bacteriophages are composed of a head, containing the nucleic acid, and a tail; they resemble sperm. At the base of the tail, the virus-specific attachment organs secure the virus to the cell wall of a sensitive bacterium with receptors. The tail then contracts, inserting a protein tube into the bacterium, just as a needle and a syringe act to inject a substance. The bacteriophage's nucleic acid then moves from the head through the tube into the bacterium—effecting attachment, penetration, and uncoating in a single step.

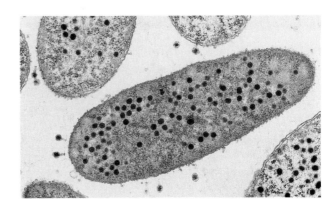

EM of a thin section of *E. coli* bacteria infected by bacteriophage T4.

Strategies of Viral Multiplication

The early events during any virus infection use the cellular signals and machinery to deliver the viral chromosome into the cell. Some viruses have evolved elaborate mechanisms to accomplish this, but the end result is always the same: access to the intracellular environment so as to reproduce more viruses. When a viral chromosome composed of either DNA or RNA enters a cell, it must accomplish two things: (1) the expression of new viral proteins in some temporal order and (2) the replication or duplication of the viral nucleic acid. The various strategies for accomplishing this are different for different virus groups, so the detailed descriptions are best left for later chapters. The expression of the viral genetic information relies on a series of steps that translate a genetic code stored chemically in the sequence of DNA or RNA nucleotides. The translation takes the chemical form of the synthesis of a protein that is composed of a defined order of amino acids. DNA (or RNA) stores the information; proteins act by carrying out the functions required for replication of the DNA (or RNA).

Assembly, Maturation, and Egress

Animal viruses have evolved several different methods for their assembly in and release from in-

fected cells. With many viruses, the viral nucleic acid is packaged with capsomeres in the nucleus or the cytoplasm, and the dying cell releases the virus particles as it disintegrates. In enveloped viruses, however, viral structural proteins or glycoproteins (proteins that have carbohydrates associated with them) are inserted through the plasma or other membranes of a cell. The viral chromosome-protein complex that has been assembled in the host cell then migrates to the position of the viral glycoprotein at the inner or cytoplasmic side of the cellular membranes and associates with it. This promotes the formation of buds where the cellular membrane containing the viral glycoproteins surrounds the nucleocapsid core, and the virus particle is extruded into the extracellular environment. Thus, the maturation and cellular release steps are coordinated, and virus production goes on for hours as a continuous release process proceeding at the cell surface.

Finally, some bacteriophages encode the information to synthesize a protein that actively promotes lysis (dissolution or disintegration) of a host cell. It is, of course, critical for the virus to withhold the synthesis of this protein, or to block its function, until very late in its replication cycle. Active lysis mechanisms, while efficient in spreading infection, can be lethal for the virus if the temporal events are not controlled properly (they are always

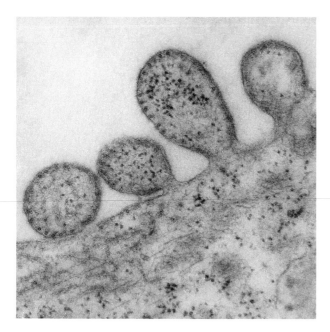

EM of measles virus particles at different stages of budding from an infected host cell; the small circles at the edges of the buds are cross sections through virus nucleocapsids.

ruses be included in this definition? Volumes have been written to elucidate the properties of a living thing, but consensus has been rare in these debates.

Viruses employ a common genetic code (the sequence of nucleotides in DNA and RNA that determines which amino acid is found at a specific position in a protein) to store information for their development and replication; indeed, the code is universal to all reproducing organisms. This could suggest a single origin event—or, alternatively, the evolution of the best-of-all-possible genetic codes from many origin events. Viruses must, of course, share the same code and signals as their host cells, in order to duplicate themselves within those cells.

Viruses, as we have seen, can program their own replication within the confines of a cell. They have a plan, satisfying Aristotle's definition of life, encoded in the format of all living things. And, like other living things, viruses evolve and respond to environmental changes. Whether this describes the simplest of living forms or an extraordinarily complex combination of nucleic acids and proteins that are merely chemicals depends upon one's view of life.

lethal for the host cell). Premature cell lysis, before infectious virus particles are produced, would result in no progeny virus and loss of the species.

It is clear, then, that viruses have evolved to reproduce themselves in a remarkably diverse set of ways. As Chapters 4 through 9 will describe in detail, viruses have evolved life styles that are unique, in that some have even left the standard paradigm of information flow in life—DNA to RNA to protein. Yet viruses are dependent upon their host cells. They must obey the rules and laws of life in a cell.

Are Viruses Alive?

The problem with this question is the same one that faced Robert Koch in defining a causative agent. How do we define life? And, once we do, can vi-

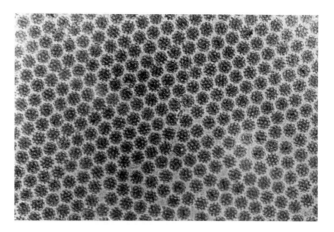

The symmetry of the Semliki Forest virus particles here shows how crystals of viruses can form (the sample is frozen, which orders the particles).

In the last chapter of this book, it will become clear that viruses, as we have now defined them, are part of a continuum of genetic elements—viruses, viroids-virusoids, plasmids, transposable elements, and insertion elements—that leads to ever-simpler forms of life: finally, to a naked DNA molecule that contains only the most primitive bits of information. We will see that it is possible that these simplest of elements could be part of a continuing creation of new life forms within a cell. (Is this a revival of the theory of spontaneous generation?) If they do play such a part, then viruses—with their ability to move between cells, between hosts, or even, in some cases, between species—will contribute to the evolutionary changes of the host by bringing it new genetic information. There is good evidence that the vestiges of viral DNA are present in the chromosomes of human beings, passed on from generation to generation as passengers in our genetic endowment. At present it is not clear whether this viral DNA results from evolutionary accident or is indeed a fellow traveler selected for a value we do not yet understand. Viruses, however, do have at least two clear roles in our evolution. First, the diseases they cause select for resistant or less sensitive hosts; the second role follows from the direct incorporation of viral genetic information into our own lineage. Clearly, viruses are intimately associated with our lives. If they are not themselves alive, they live with us.

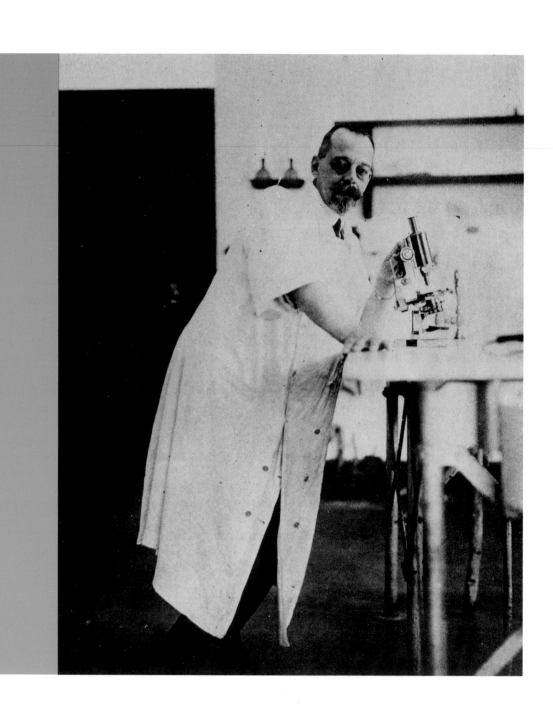

The Bacteriophages

Gradually there is coming into being a new branch of science—molecular biology—which is beginning to uncover many of the secrets concerning the ultimate units of the living cell.

Warren Weaver, 1938

Between 1915 and 1970, the viruses of bacteria were to play a central role in shaping the concepts of modern virology. But, more than that, the bacteria and their viruses became the focus of an intense research effort that led to a revolution in the biological sciences and laid the foundations for an entirely new approach that came to be called molecular biology. Out of this research and its new and powerful tools came a clear understanding of life processes at the molecular level. For the first time, we gained a fundamental understanding of genetic information—how it is stored, how it is used, and even how to change it.

Molecular biologists are now able to isolate, adapt, and express this genetic information, even in foreign hosts. They can analyze the genetic programs of an organism at the level of the molecules that carry out its cycle of life. It is possible to identify faulty genes, and in the future it may be feasible to correct these defects and even to alter the evolution of life on Earth. These new insights have the power to change us; we need the wisdom to act or not to act upon them.

Left: Felix d'Hérelle, who first isolated viruses of bacteria and named them bacteriophages.

Above: EM of T4 bacteriophages.

Historical Perspectives

Concepts and capabilities of this magnitude often have modest beginnings, and so it was with molecular biology. In 1915, Frederick W. Twort, the superintendent of the Brown Institution of London, was in search of viruses, not too fastidious, that could grow on simple artificial media in a laboratory. Twort knew that true viruses could replicate only inside a cell, but he wondered if the evolutionary predecessors of these viruses might be able to duplicate themselves outside a host. To test this, he inoculated a dish of nutrient agar with the smallpox-vaccine (vaccinia) virus, in the hope of finding a way to replicate this agent or its relatives without resorting to animals. Although this experiment failed to replicate the vaccinia virus, bacterial contaminants grew in the agar dish very readily.

As Twort waited for the virus to replicate, he noticed that the bacterial colony, containing millions of organisms, underwent a visible change. The bacteria became ''watery looking'' (more transparent), and they were no longer able to replicate; they died. Twort called this phenomenon ''glassy transformation,'' and he went on to show that infecting a healthy, normal bacterial colony with a trace of the glassy transforming principle would kill the bacteria. This glassy material readily passed through the finest porcelain filters. It could be diluted a millionfold and placed on fresh bacteria, and it would regain its strength (replicate). The agent could be stored for six months but lost its activity if heated.

Twort published a short note about this in 1915, suggesting that one explanation of his experiments was that a virus of bacteria, capable of killing them, had been identified—it was Twort's only publication on the topic. His research was interrupted by World War I, in which he served, and he did not continue this line of research after the war. In 1944 his laboratory was destroyed during a bombing raid. Twort died shortly thereafter, but his contribution will be remembered.

Meanwhile, in August 1915, Felix d'Hérelle, a Canadian medical bacteriologist working at the Pasteur Institute in Paris, began to investigate an epidemic of dysentery that was rampant in a cavalry squadron quartered just outside Paris, in Maisons-Lafitte. This disease was caused by a bacterium called the Shiga dysentery bacillus (*bacillus* describes the rod shape of these bacteria, first seen by Leeuwenhoek), which d'Hérelle proceeded to detect in filtered emulsions of the feces of sick men. He then spread these bacteria on agar culture dishes to permit them to replicate and to isolate them. D'Hérelle was careful to record that while the bacteria grew and covered the surface of the culture dish, he sometimes saw clear, circular spots where no bacteria grew; he called these spots *taches vierges* (plaques). D'Hérelle decided to follow the isolation of the bacteria and the clear plaques from

Frederick W. Twort

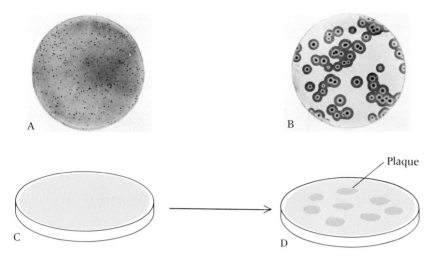

Lawn of bacteria covering the surface of a petri dish (C). Each plaque (A, B, D) arises when one infectious virus kills a bacterium and infects the neighboring bacteria. Plaques made by different viruses often have very different morphologies.

a single patient, using samples taken during each day of illness, to see if the changes in the disease correlated with changes in the plaques of bacteria. Each day, for four days, he began with a sample of feces from the same sick man, cultured the bacteria, and looked for plaques. For three days d'Hérelle found good bacterial growth on all his culture dishes and no trace of a plaque on any of them. Then, he reported:

> The fourth day, as on the preceding days, I made an emulsion with a few drops of the still bloody stools, and filtered it through a Chamberland candle; to a broth culture of the dysentery bacillus isolated the first day, I added a drop of the filtrate; then I spread a drop of this mixture on agar. I placed the tube of broth culture and the agar plate in an incubator at 37°C. It was the end of the afternoon, in what was then the mortuary, where I had my laboratory.
>
> The next morning, on opening the incubator, I experienced one of those rare moments of intense emotion which reward the research worker for all his pains: at the first glance I saw that the broth culture, which the night before had been very turbid, was perfectly clear; all the bacteria had vanished, they had dissolved away like sugar in water. As for the agar spread, it

was devoid of all growth and what caused my emotion was that in a flash I had understood: what caused my clear spots was in fact an invisible microbe, a filterable virus but a virus parasitic on bacteria.

> Another thought came to me also: "If this is true, the same thing has probably occurred during the night in the sick man, who yesterday was in a serious condition. In his intestine, as in my test tube, the dysentery bacilli will have dissolved away under the action of their parasite. He should now be cured." I dashed to the hospital. In fact, during the night, his general condition had greatly improved and convalescence was beginning.

In 1917 d'Hérelle published these observations and others. He described a general procedure to isolate viruses of bacteria, using a plaque assay (assessing clear spots on a lawn of bacteria) to quantitate their strength. D'Hérelle named this group of viruses bacteriophages, or bacteria-eating agents. These publications contain no reference to the work of Twort; d'Hérelle knew nothing about "glassy transformation" or the previous research in this area. Generally, both men share the credit for these early observations and the discovery of bacte-

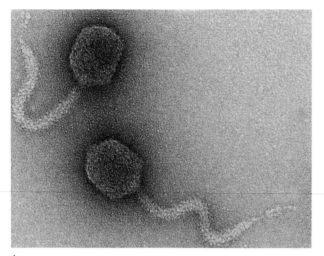

A

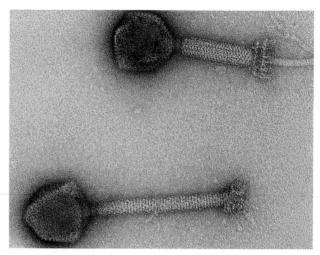

B

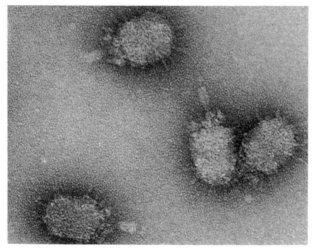

C

Bacteriophages vary in size and morphology. A: Phage γ has an elongated icosahedral head (55 nm dia.) and a flexible, noncontractile tail. B: Phage SP105 has a spherical icosahedral head (65 nm dia.) and a contractile tail. C: Phage φ29, with an extended icosahedral head (35 nm dia.), shows a neck with two collars and twelve appendages, plus a short tail.

rial viruses. D'Hérelle, however, continued his investigations in two distinct directions.

From the beginning he was enormously impressed by his first experience with the dysentery patient. D'Hérelle felt that the disease could be described in two phases. First, the Shiga bacillus infected a patient and, as it replicated, caused disease symptoms. Many people recovered over the course of a week because the bacteriophage then attacked the high concentrations of bacteria, and the dysen-

tery was cured. By 1918, d'Hérelle had thirty-four case studies of dysentery that seemed to fit this pattern. Moreover, bacteriophages were rapidly isolated from a wide variety of bacteria that cause cholera, anthrax, diphtheria, bubonic plague, and many other illnesses. D'Hérelle now felt sure that his observations were general, applicable to all diseases caused by bacteria. It was only a step to the suggestion that bacteriophages (a term often shortened to phages) should be actively administered to patients to protect them from such diseases.

D'Hérelle actively proselytized for this therapeutic approach, and a great deal of medical research was carried out during the 1920s to test this new hypothesis. (These ideas were so popular that Sinclair Lewis, writing with real scientific insight in the 1920s, used medical research on bacteriophages as a main plot element in his novel *Arrow-*

smith. The field trials of bacteriophage therapy during a fictional outbreak of bubonic plague in the West Indies introduce the reader to the problem of who should serve as a control group, getting no treatment while a real therapeutic possibility exists. Can one withhold a therapy that might save a life? This is a question that has revisited all of us during the AIDS epidemic of the 1980s and 1990s.) Despite hundreds of reports in the literature over a twenty-year time span (1918 to 1938), bacteriophage therapy failed to show efficacy in curing diseases. The reasons for this failure remain unclear.

Even though one of d'Hérelle's approaches took him down a blind alley, his other investigations into the basic nature of bacterial viruses opened new doors. He developed the fundamental method for quantitating viruses: the plaque assay. If one begins with a solution of a million viruses, one can make a series of dilutions to obtain test tubes with 1000, 100, 10, and 1 virus particle(s) per tube; that is, when mixed with bacteria and spread out on the surface of culture dishes, these solutions yield plates with 1000, 100, 10, and 1 plaque(s) on them. One virus makes one plaque, spreading from the first infected bacterium in an ever-widening circle. The progeny of one parental bacteriophage populate each plaque.

D'Hérelle went on to show that the virus must first attach to a bacterium, an attachment that often determines the specificity of its host and its killing abilities. The bacteriophage that kills the Shiga bacillus, for example, does not attach to the staphylococcal bacteria. Once the virus is attached and inside a bacterial cell, a latent period ensues during which no visible change is apparent under a light microscope. The bacterial culture will then suddenly lyse, breaking open the cells and clearing the turbid suspension of bacteria. An increased number of viruses are released into the medium to attach and kill again. D'Hérelle was the first to describe this cycle of replication.

Yet all was not so simple in the real world of living organisms. Sometimes these viruses seemed to appear from nowhere. A bacterial culture that did not contain detectable virus (made no plaques) would sometimes—presumably in response to an unknown, outside influence—generate plaques on a petri dish. The bacteria could be grown in pure culture—that is, separated from all other organisms—and yet still somehow give rise to bacteriophages. It appeared that some viruses had a profoundly intimate relationship with their hosts; in these cases, there simply was no way to separate the host cell and its virus, which could periodically give rise to cell lysis and virus production. This so-called lysogenic state led to a good deal of confusion and to further experimentation about the fundamental nature of this close association.

The Beginning of Molecular Biology and the Phage Group

By the last few years of the 1930s, conditions were right for the beginnings of a new era. Throughout the 1920s and the 1930s, scientific research was funded and supported by a few foundations and some government agencies. It was not an organized, systematic arrangement, but rather a constant struggle for funds to run a laboratory. One of the real leaders during this period was the Rockefeller Foundation, which had representatives and offices throughout the world, identifying the best scientists and making sure they were supported. Its motto, "Make the peaks grow higher," symbolized its philosophy of supporting only the best in chemistry, physics, biology, and the medical sciences; its early support of research at Oxford University into the ability of penicillin to cure bacterial diseases changed the world—and made everyone forget bacteriophage therapy.

In 1938 Warren Weaver, the director of the Rockefeller Foundation, noted the emergence of a new branch of science in his annual report to the trustees (see the epigraph to this chapter). He continued, "Among the studies to which the Foundation is giving support is a series in a relatively new field, which may be called molecular biology, in which delicate modern techniques are being used to investigate ever more minute details of certain life processes.'' What Weaver had recognized were the beginnings of three very different approaches or schools of thought in biology: the structural school, the informational school, and the biochemical school.

The structural school was based, in the main, at Cambridge University in the Cavendish Laboratory, where Max Perutz and his colleagues were developing methods to determine the three-dimensional structures of proteins such as hemoglobin (which carries oxygen in the blood) and chymotrypsin (which helps to digest and break down proteins in food). As we have noted, the proteins of living organisms are linear series of twenty different amino-acid subunits, any one of which can be found in any given position of a protein polymer; proteins differ because their amino-acid sequences are different. For example, hemoglobin is composed of two different proteins, called the alpha chain and the beta chain; the former polymer has

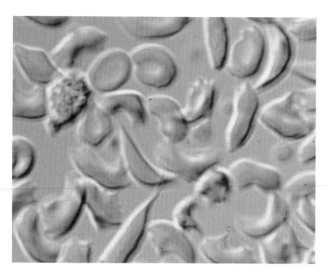

''Sickle'' red blood cells.

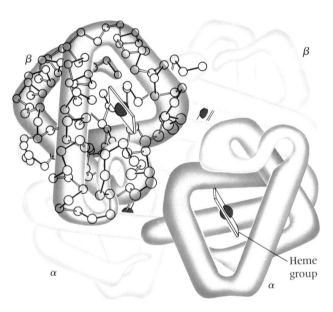

Conformation and shape of the hemoglobin molecule: two alpha chains and two beta chains fold, producing a pocket that holds a heme group.

141 amino acids and the latter 146. At each position in the protein (1 through 141 or 146, respectively), a specific amino acid (one out of the twenty) resides. The sequence of amino acids in a protein, which is termed the primary structure, determines the chemical properties of the protein; in particular, it guides the protein to fold into a three-dimensional shape, which determines the protein's function.

A hemoglobin molecule is composed of two alpha chains and two beta chains: it is therefore a tetramer assembled from these four proteins. They form a pocket to hold a deep red, iron-containing heme molecule, which binds either oxygen or carbon dioxide. Oxygen is taken up in the lungs, carried to the tissues of the body, and exchanged for carbon dioxide, which is given off again in the lungs.

This whole process becomes defective in certain disease states, such as sickle-cell anemia. Individuals with sickle-cell disease have the wrong amino acid in position 6 (out of 146 amino-acid subunits) in the beta chains of the hemoglobin protein. The new amino acid in that position results in the improper folding of the protein (so that the red blood cell assumes an altered or sickle shape) and in poor protein function, which diminishes the supply of oxygen to the tissues. Because sickle-cell disease is inherited, the information in the chromosomes must somehow determine both the specific amino acid and its position in the protein. It is clear, then, why some scientists want to study the structure of proteins and why Max Perutz, Linus Pauling, and others of the structural school helped to lead the way to new concepts.

The second approach to molecular biology, termed the informational school, aimed to understand how genetic information determines the structure and function of proteins, and a major tool of this group was genetics. Max Delbrück, who became one of the founders of the informational school, was trained in his native Germany at the University of Göttingen as a physicist; his first posi-

Max Delbrück.

tion was at the Kaiser Wilhelm Institute for chemistry in Berlin, where a group was actively discussing the application of the new ideas in quantum physics to heredity in living organisms. Delbrück developed a "quantum mechanical" model of the gene. In 1937 he went to Caltech as a research fellow. His work there took two paths: participation in the structural group with Pauling and an introduction to the bacteriophages with Emory Ellis. Ellis was studying a group of phages (they came to be called the T phages—T2, T4, and so on) because he believed that they represented a simple model for understanding cancer viruses and even for understanding how a sperm fertilizes an egg, resulting in the development of a new organism. Delbrück appreciated that this model system had many experimental advantages for getting at the questions: What is the nature of genetic information? How does it function?

At the outbreak of World War II, Delbrück remained in the United States (at Vanderbilt University) and was joined by an Italian refugee, Salvador E. Luria, who had worked at the Pasteur Institute studying bacteriophages with Eugene Wollman, but who escaped to America as the war began. They met in late December 1940 and went off to Luria's laboratory at Columbia University to focus their attention on the bacteriophages. These two scientists were to recruit and lead a growing group of researchers focused on bacterial viruses as a model for understanding life processes at the molecular level. Crucial to the success of that venture was an invitation to spend the summer of 1941 at the Cold Spring Harbor Laboratory doing experiments. The German physicist and the Italian geneticist joined forces throughout the years of World War II to travel about the United States and convince a new generation of biologists that they should join "the phage group."

Soon Tom Anderson, an electron microscopist who was a fellow at the Radio Corporation of America (it had one of the only electron microscopes in the United States), met Delbrück and agreed to help him see what bacteriophages looked like. Luria and Delbrück had to get special clearance from the government to visit the RCA facility in Princeton, New Jersey, but by March 1942, they obtained the first clear pictures of these viruses, remarkably demonstrating that they had heads and tails. At about this time, the first mutant forms of these bacteriophages were isolated by Luria, and analysis of their genetics began with a publication in 1945. By 1946 Delbrück was teaching the first phage course at Cold Spring Harbor, which brought new scientists into the field. In March 1947, the first phage meeting attracted eight people—and the informational school was born.

The third school used biochemistry to study bacteriophages. Early on this group included Seymour Cohen, who took the first phage course in 1946 and trained with Erwin Chargaff at Columbia University. Later, Arthur Kornberg and others worked out the biochemical pathways by which genetic information is synthesized in bacteria and their viruses. This group analyzed the chemical makeup of phages and their genetic information, elucidating the chemical reactions in a cell that result in the duplication of these viruses. All three schools—the structural, informational, and biochemical—were to lay foundations for the coming revolution.

In 1944 Erwin Schrödinger, the great theoretical physicist, published his book *What Is Life?* The central problem that it pointed out was the clear ability of life to reproduce itself faithfully, implying a mechanism of information transfer from generation to generation that is remarkably reliable. How, Schrödinger wondered, could the chemical that encodes that information be so stable at 98°F, despite the destabilizing bombardment of thermal energy? Although the answer to the problem is that the information is not in fact stable (mutations occur frequently and are not all bad), his book continued to attract to biology excellent scientists from the ranks of physics.

That same year, O. T. Avery, C. Macleod, and M. McCarty of Rockefeller University showed that DNA taken from one bacterium and added to another bacterium with a different set of genetic traits could transmit an inherited change to the recipient organism. In a convincing set of experiments (a combination of the biochemical approach and the informational-genetic school), they showed that the molecule in a bacterial cell that stores the genetic information is indeed DNA. Although the implications of these experiments were not fully realized by the scientific community at the time, this work did focus some laboratories on the study of DNA molecules.

DNA, as you will recall, is a polymer composed of monomeric subunits called nucleotides that are linked to one another in a linear fashion. Nucleotides are chemically composed of three parts, a base that is linked to a sugar (in DNA this is always deoxyribose), which in turn is bound to a phosphate group. The nucleotides are linked in a bonding pattern of sugar-phosphate-sugar-phosphate-sugar-, and so on, called a phosphodiester bond; the bases attached to the sugars are free to interact with other bases. DNA has four bases: adenine, thymidine, guanine, and cytosine—commonly abbreviated as A, T, G, and C. Thus, four different nucleotides positioned in a linear polymer at specific places produce a sequence—ATTGCCA, for example. DNA has four nucleotide monomers linked in a polymeric sequence, while proteins have twenty amino-acid monomers, also in a linear polymeric sequence. In both cases the sequence is important: which nucleotide or amino acid is in which position in the polymer makes all the difference. (The other major nucleic acid in a cell is RNA. In RNA there are also four bases: adenine, uracil—which takes the place of thymidine—guanine, and cytosine. Each base is linked to a sugar—ribose in this case—which in turn is linked to a phosphate group, making a nucleotide; these ribonucleotide monomers are polymerized into RNA.)

Erwin Chargaff at Columbia led one of the first groups to quantitate the amount of each nucleotide—A, T, G, and C—in DNAs from various organisms. He found a curious fact: in DNA the amount of T always equaled the amount of A, while the concentration of G was always equal to that of C. Chargaff's rules—A = T and G = C—were to be crucial in a few years; at the time, every-

The Cold Spring Harbor phage group, 1952. Martha Chase and Al Hershey are second and third from the left.

one agreed they were important but no one could quite say why.

By 1952, the phage group at Cold Spring Harbor, led by Martha Chase and Al Hershey, carried out a critical demonstration that has come to be called the "Waring blender experiment." Hershey and Chase realized that the bacteriophages were composed of two kinds of chemicals, an outer coat of protein and an inner core (in the head) of DNA. These two parts could each be labeled, using radio-isotopes in amino acids for the proteins (the sulfur atoms were labeled) and in phosphate for the DNA (the phosphate atoms were labeled). When a radio-actively labeled virus is mixed with a bacterium, the protein coat (sulfur label) and DNA (phosphate label) both attach to the bacterium and stick tightly. Shortly after attachment, the DNA leaves the phage and enters the bacterium. The protein coat remains outside the cell throughout the infection period. Hershey and Chase showed this by placing these virus-bacterial complexes in a blender. The shear forces strip off the protein coats of the viruses from the bacteria without hurting the bacterial hosts. Moreover, in thirty minutes each bacterium lyses and liberates a hundred new phages, even though the viral protein coat is no longer attached to the bacterium. Clearly, the information needed to make the hundred new phages resides in the DNA and not in the protein coat. DNA, therefore, contains the genetic information—confirming the earlier experiments of Avery and his colleagues. This was just the right time to bring this observation to the attention of the growing scientific community of phage workers.

In 1950 James D. Watson, working with Luria and the phage group, received his doctorate from

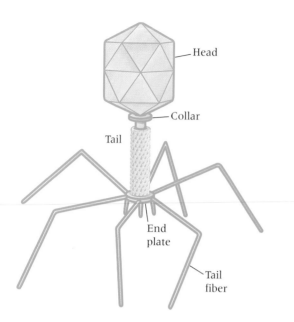

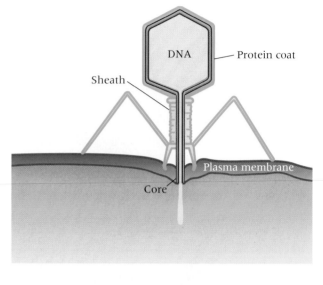

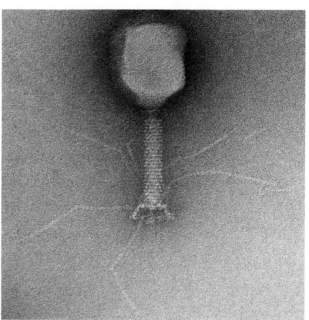

Above: Diagram of bacteriophage T4 (left); T4 attached to a bacterium and injecting its DNA (right). The tail contracts, forcing the injection tube through the wall and into the cytoplasm; the viral DNA travels from the head into the cell, while the coat proteins remain outside.

Right: EM of bacteriophage T4. The tail, which has a helical rod symmetry, is used as a syringe to inject the viral DNA contained in the head, which approximates an elongated icosahedron. The fibers that project from the end of the tail are proteins that attach the virus to a bacterial host.

the University of Indiana. By 1951 he had joined the Cavendish Laboratory at Cambridge (the structural school) and teamed up with Francis H. C. Crick, who was using X-ray crystallography to study protein structure. Using an X-ray analysis of

DNA structure carried out by Rosalind Franklin and Maurice Wilkins, Watson and Crick were able to build a model of the structure of DNA that explained how Chargaff's rules came about and why, in all DNA molecules, A = T and G = C.

Few discoveries in science have had both the immediate and the far-reaching impact of the elucidation of the structure of DNA, the genetic material. There are several reasons for this. One molecule of DNA is made up of two polymerized nucleotide chains, which wind about each other to form a double helix. The phosphate groups of the monomeric subunits are on the outside of the helix, while the bases are on the inside. What holds one strand to the other are chemically attractive forces called hydrogen bonds, involving charge attractions between hydrogen atoms and nitrogen or oxygen atoms of the bases. Where one strand of DNA has a T, it is attracted by (pairs with) an A on the opposite helix. Where one strand has a C, it is always opposite a G. So A = T and G = C in any double-stranded DNA polymer.

These specific pairing rules also explain how DNA can replicate itself. If the strands separate, then the sequence on one strand of DNA—

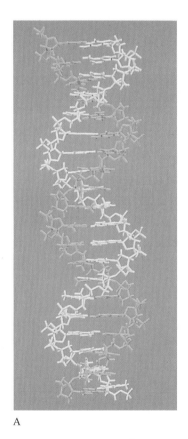

A

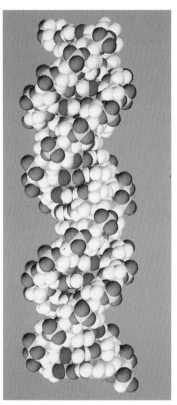

B

Computer-graphic representations of DNA structure. A: A stick model showing two turns of the helix. The bases are on the inside of the helix (like bridges) where hydrogen bonds—A to T or G to C—hold the two chains together. B: A space-filling model of the double helix. Spherical balls more accurately represent the space occupied by the different atoms and their positions in a DNA molecule, but they do not show the base pairing or other features as clearly.

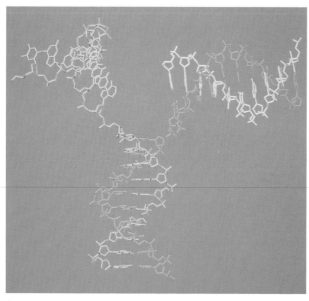
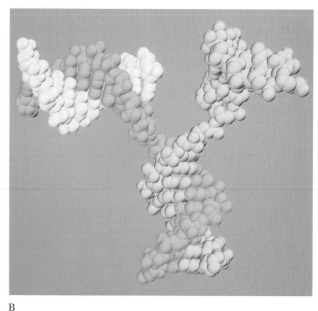

A   B

Computer-graphic representations of replicating DNA; stick (A) and space-filling (B) models. The two strands of the double helix separate and the nucleotide sequence of each is used as a template to synthesize a new, complementary copy of the strand.

ATTGCGA, for example—predicts by the pairing rules what the complementary sequence on the other DNA polymeric strand must be—in this example, TAACGCT. In other words, one strand reproduces its sequence in a complementary fashion on the other DNA strand. When the two polymeric chains of DNA in a helix separate, each is able to act as a template, guiding the synthesis of a complementary sequence of nucleotides that always follows the rules A = T and G = C (A pairs with T and G pairs with C).

Not only did this double-helix model provide the structure of DNA, it predicted a mechanism for the faithful duplication (via strand separation) of the genetic information. By 1958, M. Messelson and F. Stahl at Caltech had carried out an experiment proving that DNA indeed replicates itself by separating the two parental polymeric strands and pairing with two new daughter strands. Beyond

this, the Watson and Crick model offered for the first time an understanding of how information, in this case genetic information, could be stored in a chemical molecule.

In a long polymer, the nucleotide sequence can encode information. That sequence means something to both the virus and the host cell. Somewhere in the order of nucleotides is a genetic code that holds the information to reproduce hundreds of viruses. The genetic code employs an alphabet of only four letters—A, T, G, C—but by using these letters in a variety of specific orders, it can achieve enormously complex messages. For example, some simple viruses have chromosomes with five thousand nucleotides in a specific sequence. Because each position can have any of four nucleotides, the total number of possible nucleotide sequences for this virus is 4^{5000} (four multiplied by itself five thousand times). This number is equal to

1.995063116880758 times 10^{3010}. To print out this number would take a page and a half of single-spaced text. Because of this exponential function, even simple viruses can have great diversity.

How does the information stored in DNA get translated into viral and cellular function? One aspect of the answer is that the sequence of nucleotides in DNA determines the sequence of amino acids in proteins—a sequence that, as we saw earlier in our introduction to the structural school, determines the folding and, hence, the function of a protein. Hemoglobin was our example. There are two genes for hemoglobin, the alpha gene and the beta gene, each composed of a nucleotide sequence that determines the sequence of amino acids in the alpha and the beta protein chains, respectively. A change in a specific nucleotide in the beta gene (a mutation) changes an amino acid in the beta protein chain and causes the protein to function poorly.

From 1955 to 1970, the way in which proteins are synthesized in a cell was elucidated. The nucle-otide sequence from one of the two strands of DNA is first copied into an RNA molecule, called messenger RNA (mRNA). This process, called transcription, uses the A = U (RNA has U in place of T), G = C pairing rules, so the mRNA is a faithful but complementary copy of one DNA strand. The mRNA then directs the synthesis of proteins (on cell structures called ribosomes) by assembling amino acids in a specific order dictated by the sequence of nucleotides. Just which sequences code for which of the twenty different amino acids was finally worked out by M. Nirenberg, S. Ochoa, G. Khorana, and others. The genetic code was broken; the syntax of this language was understood. Throughout this period, the study of viruses continued to play a key role, first in elucidating the existence of mRNA (with the T2 phage in 1956) and then, by means of phage genetics, in the discovery and confirmation of biochemical approaches to deciphering the genetic code.

An example may make this process more concrete. If a nucleotide sequence in the DNA polymer

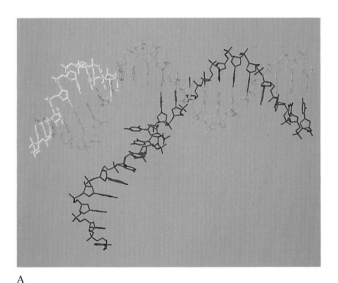

A

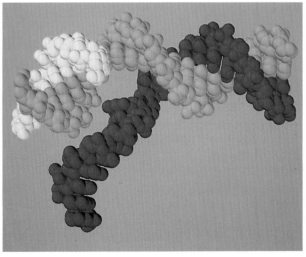

B

Computer-graphic representations (A: stick, B: space-filling) of the process of transcription. One of the two strands of DNA is copied into RNA, using the A = U and G = C pairing rules.

happens to be ATG-AAA, then opposite these nucleotides in the double helix is a complementary sequence, TAC-TTT (A = T, G = C are the pairing rules). This double-stranded DNA uses the same pairing rules to transcribe an mRNA copy of one strand of the DNA: AUG-AAA (U replacing T in RNA). Once the mRNA has migrated out of the nucleus and joined a ribosome in the cytoplasm, the nucleic-acid code AUG-AAA is translated by the mRNA into an amino-acid code producing proteins of defined amino-acid sequence; in point of fact, AUG-AAA will direct methionine-lysine (two of the twenty amino acids) into a protein polymer in that order.

It should be emphasized that there are two complementary sequences of nucleotides in DNA—in our example, ATG-AAA and TAC-TTT. The sequence that is identical to the mRNA (AUG-AAA), which is translated into an amino-acid sequence, is called the positive sequence, the sense sequence, or the plus-strand sequence; the complementary sequence is called the negative sequence, the antisense sequence, or the minus-strand sequence. In almost all cases, only the sense-strand sequence of nucleotides will direct the synthesis of a protein that has a function; that is, only one DNA strand has the code that makes sense.

(These plus and minus sequences have special relevance for the classification and life histories of viruses, as later chapters will clarify in detail. There is a group of viruses whose genetic information is stored not in DNA but in positive or sense RNA, like the mRNA; a different set of viruses uses a negative or antisense RNA molecule for the same purpose. As we will see in Chapters 6 and 7, moreover, the retroviruses have a positive, sense RNA that is copied into a double strand of DNA by an enzyme in the retrovirus particle called reverse transcriptase. This DNA is then transcribed into mRNA that is the same polarity as the genomic RNA of the retrovirus, so it is packaged into retrovirus particles directly.)

The description of the structure of DNA helped to define the experiments of the next fifteen years, which reconfirmed the importance of this discovery. Furthermore, the three approaches to studying life processes—structural, informational-genetic, and biochemical—each made critical and interactive contributions to our understanding of how living organisms replicate themselves. It was Delbrück and Luria who first chose and then campaigned for a simple experimental system, the T phages. The phage group of scientists, eight strong in 1947, had grown in one generation to hundreds of scientists publishing thousands of papers describing minutely the replication of a simple virus.

The Synthesis of T2, T4 Bacteriophages

In 1946 Seymour Cohen (a student of Chargaff and a member of the biochemical school) took the phage course at Cold Spring Harbor with Max Delbrück. He then began a study of the T2, T4 phage DNA and how it was synthesized. (T2 and T4 bacteriophages are so closely related that for the purposes of our discussion they can be considered

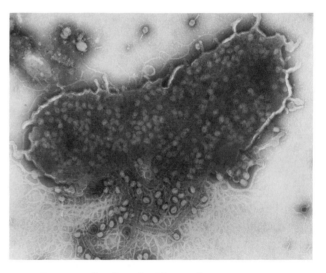

Bacterium *E. coli* infected with T4 phages.

A. *E. Coli* DNA

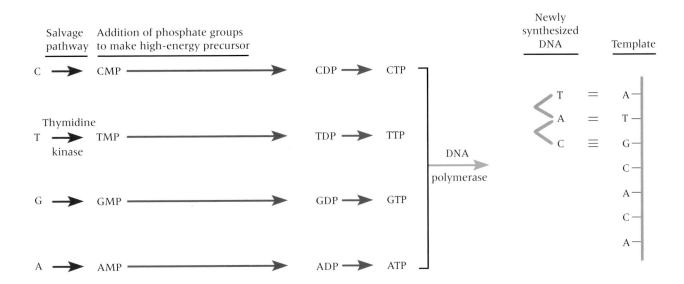

B. T4 DNA

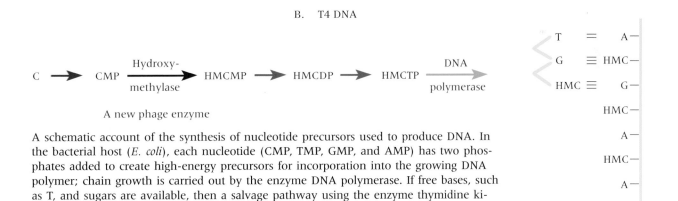

A schematic account of the synthesis of nucleotide precursors used to produce DNA. In the bacterial host (*E. coli*), each nucleotide (CMP, TMP, GMP, and AMP) has two phosphates added to create high-energy precursors for incorporation into the growing DNA polymer; chain growth is carried out by the enzyme DNA polymerase. If free bases, such as T, and sugars are available, then a salvage pathway using the enzyme thymidine kinase converts T to TMP. T4 DNA does not have C, but instead substitutes hydroxymethylcytosine (HMC), which pairs with G. The bacterial cell has no way to convert CMP to HMCMP (hydroxymethylcytosinemonophosphate). The enzyme that does this is encoded by a gene not in the bacterium, but in the T4 chromosome. This gene directs the synthesis of an enzyme, hydroxymethylase, that converts CMP to HMCMP. The virus donates new genetic information to the infected cell.

identical.) By 1952, his research group reported a curious observation: the four nucleotide bases in the T2, T4 phages are not A, T, G, and C, but A, T, G, and hydroxymethylcytosine (HMC). The host bacterium contains A, T, G, and C in its DNA, but the virus modifies or chemically alters *its* DNA by using HMC instead of C. The biosynthesis of the four nucleotide precursors for DNA occurs in a number of discrete steps in a cell, as outlined in the figure on page 39.

Cohen appreciated that the bacterial host for the T2 phage, *E. coli,* did not have the biosynthetic capacity to synthesize HMC; it could only produce C. He went on to show that the ability to synthesize HMC was acquired within a few minutes after the phage injects its viral DNA into the host cell. The fact that uninfected bacteria cannot synthesize HMC but virus-infected cells can was the first clear indication that a viral-encoded gene could contain the information to produce a new protein—in this case an enzyme that synthesizes a precursor for DNA synthesis, HMC. The virus brings in new genetic information and reprograms the events of the host cell. That concept fit with the Hershey-Chase blender experiment and pointed the way to the search for viral-encoded gene functions.

Cohen's experiment, moreover, shed new light on the so-called latent period of virus infection. Just before the war, Ellis and Delbrück had described what they called a "one-step growth curve" for bacteriophage infection. The one-step growth curve with T2 showed viruses attaching to cells, loss of infectivity (no plaques), and thirty minutes later a burst of new phages released from each cell. Just what was happening inside a cell from two minutes to thirty minutes, the latent period, was now under intensive study.

In fact, a whole series of genes encoded by T2 are expressed at defined times during the latent period. From one minute after infection, when the viral DNA has been injected into the host, a series of mRNAs is produced, copying a set of genes in the viral DNA. These mRNAs in turn synthesize proteins whose function it is to manufacture nucleotides—the precursors of DNA—and to polymerize these nucleotides, replicating the viral DNA. These are the early gene functions, or early proteins: they are produced immediately, from one to seven minutes, after infection.

Other viral-encoded genes direct the synthesis of viral proteins that inhibit several normal host-cell functions. Cellular DNA synthesis is blocked immediately after infection: the bacterial DNA no longer synthesizes its mRNAs, and no bacterial proteins are produced. The T2 phage completely redirects the synthesis of proteins and DNA from bacterial to viral; only viral genes are replicated and only viral genes are expressed, thus conserving the energy of the cell and focusing it on reproducing the virus. Eventually this kills the cell.

The virus even synthesizes enzymes that degrade the cellular chromosome (DNA), and the nucleotides released are reincorporated into the viral chromosome. At seven minutes, the virus has made nucleotide precursors and added phosphate groups so as to synthesize polynucleotide chains that are complementary in sequence to the infecting viral DNA, which then replicates by separating its chains. DNA replication proceeds, producing hundreds of viral DNA copies in each infected cell.

Once that is accomplished, a new set of viral genes is used to synthesize mRNA and proteins. These genes are termed the late genes, and they direct the synthesis of structural proteins that make up the head and tail of the virus particle. The early viral proteins are, by and large, enzymes, whose function it is to catalyze chemical reactions and produce nucleotides and DNA. Enzymes speed up the rate of chemical reactions. Because a few enzyme molecules can synthesize lots of nucleotides (they are used thousands of times in repetitive reactions), the early proteins need to be synthesized only in small amounts. The late viral proteins, on the other hand, are structural and must be produced at high levels. They are used like bricks to build virus particles. Interestingly, head and tail proteins are each assembled in a separate pathway or production line, step by step by step (see adjacent figure). Then the head, with its DNA enclosed, is placed on a tail, producing a mature phage. The mRNA for late proteins is copied from hundreds of

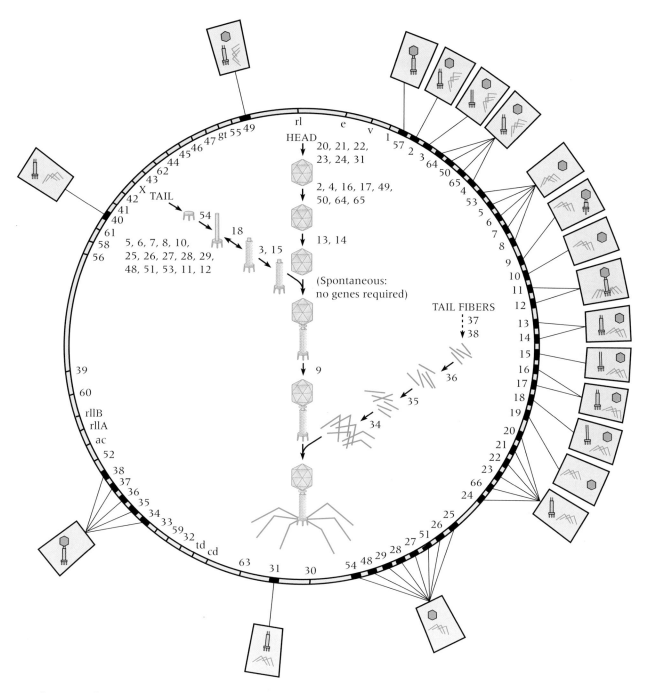

Circular map of T4 phage DNA. Each gene, assigned a letter or number, encodes one protein. Genes for particle assembly—depicted inside the ring—are matched with diagrams of the faulty structures produced when those genes are defective: for example, genes 34–38 (no tail fibers), gene 49 (no assembly), gene 9 (premature tail contraction). Mutations in genes 11–12 generate a complete but fragile bacteriophage. Numbers next to arrows in the three branches of the assembly line indicate the genes crucial to each step.

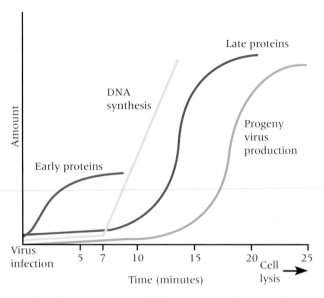

Graphical representation of the events in a bacteriophage infection. One subset of viral genes is expressed shortly after infection. These early gene products (such as hydroxymethylase) function to synthesize the nucleotide precursors of phage DNA and shut off cellular DNA, RNA, and protein synthesis. By seven minutes, viral DNA synthesis begins. Shortly thereafter, the second set of viral genes is transcribed, producing mRNA and late proteins in very large amounts to package the hundreds of viral DNA chromosomes recently synthesized.

DNA templates synthesized before late transcription, while the early mRNA is synthesized from the single parental DNA template. In part for that reason, there is more late mRNA than early mRNA, so many more late proteins than early proteins are synthesized.

The program for the replication of T2 virus in *E. coli*, therefore, is: (1) phage attachment to the bacterial cell; (2) injection of phage DNA into the host cell; (3) transcription of the early genes and synthesis of the early proteins; (4) inhibition of cellular DNA, RNA, and protein synthesis; (5) synthesis of nucleotide precursors; (6) replication of viral DNA; (7) transcription of late genes and synthesis of late proteins; (8) packaging of viral DNA and assembly of the phage head; (9) assembly of the phage tail and tail fibers; (10) assembly of complete phage particles; and (11) cell lysis and release of the progeny virus particles. This is an orderly program, where one event depends upon the completion of a prior event; you will recognize the parallel with the general sequence of viral replication reviewed in Chapter 1. Max Delbrück was correct. The whole process is much like the development of any complex organism over time. It is a good model for understanding all the life processes.

Lysogeny

Not every virus is programmed, like T4, to reproduce itself and kill the host cell. Some viruses live and replicate *with* their host cells. These viruses interact with host-cell genes and proteins, duplicating once every generation with the bacterial host. When a single bacterium has grown into a million bacteria, the viral DNA, as part of the bacterial chromosome, is present in all the million offspring of the bacterium. Only occasionally, in response to outside signals such as ultraviolet light or toxic agents, will it excise from the chromosome of the host and duplicate itself more autonomously, lysing (dissolving) and killing the host bacterium and releasing new viruses. The virus harbored by the bacterium senses the change for the worse in the environment and "bails out," killing its host and liberating progeny viruses, which now find a new home. Bacteria that show this property are called lysogenic bacteria. Lysogeny is a common phenomenon in the bacterial kingdom; it was first clearly described by André Lwoff at the Pasteur Institute in Paris shortly after World War II.

During the 1930s, however, a scientific argument was raging between d'Hérelle and his followers and Jules Bordet, who was the director of the Pasteur Institute in Brussels. D'Hérelle felt that his work clearly demonstrated the particulate nature of bacteriophages. He saw the phage as an agent that, when added from outside a host cell, entered the cell, reproduced, and killed the cell. Bordet, on the other hand, had clearly shown that the "lytic principle" originates from within the bacterium. He accomplished this by placing a single bacterium on an agar culture dish. This organism gave rise to millions of progeny: a pure culture of bacteria, with no sign of a virus. As these bacteria continued to grow, one in a thousand—or one in ten thousand—suddenly lysed open, liberating hundreds of phages that could, under the right conditions, make hundreds of plaques. It was as if the host harbored the phage in a cryptic, or hidden, form, to be called out on rare occasions. When these infected bacteria are "touched" by an outside influence, Bordet concluded, they respond by creating a dissolution factor, or lytic principle.

To put this controversy in more modern terms, the genetic information for this self-reproducing entity comes either from a virus infecting the cell or from the chromosome of the cell itself. The bacterial cell contains a set of genes that, if expressed, lyse the cell and release an infectious agent. Just which genes and chromosomes are viral and which are cellular seems to become a bit blurred here. Although neither d'Hérelle nor Bordet ever thought the other was correct; in fact both were. The idea that some viruses can place their DNA into a host cell and integrate the viral DNA into the polynucleotide chain of the host-cell chromosome was born in these early arguments.

Modern investigators have thrown light on these issues by working with the bacteriophage lambda. Lambda is a lysogenic virus of *E. coli* that has been intensively studied by the phage group over the past half-century. It represents a good example of how a virus can learn to live with its host. Lambda has a head and a tail, like T4. The head contains a double-stranded DNA molecule that is about 45,000 base pairs long. The average gene in lambda is about 1000 base pairs long (producing a protein about 330 amino acids long), so lambda virus encodes the information for about 45 proteins (one gene, one protein). In contrast, *E. coli* has a chromosome of about three million base pairs and is estimated to encode about 5000 genes and a similar number of proteins.

Lambda can attach itself to an *E. coli* host and insert its DNA into the cell. At this point, the virus enters into one of two different pathways. It may express its early viral proteins, replicate the viral DNA, synthesize coat proteins, assemble, and lyse the cell (the lytic path); or it may express a single gene and protein that keep the entire chromosome silent (no mRNAs are produced), and the infecting viral DNA then integrates into the DNA of the bacterial chromosome (the lysogenic path). Which path the virus embarks on depends on the physiology of the cell and a race between the expression of two different viral genes.

Once integrated into the chromosome of the bacterium, the lambda DNA produces mRNA from only one gene, the C_I gene. The mRNA synthesizes a protein of 236 amino acids, the repressor protein. One of the functions of the repressor protein is to recognize specific DNA sequences of nucleotides (comprising about 17 base pairs) that exist in the lambda chromosome. These key sequences are termed operator sites, and the lambda chromosome

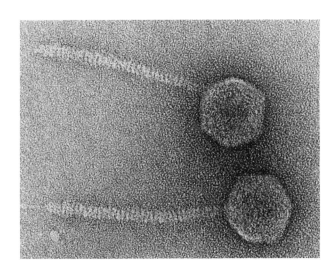

Bacteriophage lambda (λ).

has two major ones, called O_L (operator left) and O_R (operator right). When a dimer of the lambda repressor protein recognizes its operator site, it binds tightly to that specific nucleotide sequence on the viral DNA. The O_R site is located between two genes; to its left is the C_I repressor gene and to its right is a gene called cro, which produces a protein (66 amino acids long) that recognizes the same DNA sequences and binds to them. Moreover, cro is also a repressor molecule and a dimer.

At the start of every gene, just in front of the signal to begin making a protein, is a signal that is termed a promoter. A promoter is a nucleotide sequence where RNA polymerase binds and proceeds to synthesize the mRNA copy of the DNA. (RNA polymerase is the enzyme that copies DNA into mRNA by making complementary copies of the nucleotide sequence.) Every gene, therefore, needs an mRNA-start site: the promoter sequence. The C_I gene promoter is transcribed in the leftward direction, copying one of the DNA strands. The cro gene promoter is copied from the other DNA strand, in the rightward direction. In the region of DNA between these two genes (about 80 base pairs long) there are three related but distinct 17-base-pair binding sites (collectively, O_R) for both the lambda C_I repressor and the cro protein.

This intergene area, then, contains a C_I gene promoter (leftward mRNA), a cro gene promoter (rightward mRNA), and three related O_R sequences that bind both repressor proteins. Now, two of the three nucleotide sequences that make up O_R overlap the C_I gene promoter; when the cro repressor protein is bound there, RNA polymerase cannot enter at the promoter, and neither C_I mRNA nor C_I repressor protein can be synthesized. Conversely, the C_I repressor binds very tightly to the two O_R sequences that overlap the cro gene promoter, so C_I

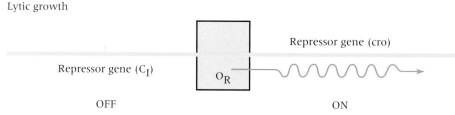

A. Prophage

Repressor gene (C_I)

O_R

Repressor gene (cro)

ON

OFF

B. Lytic growth

Repressor gene (cro)

Repressor gene (C_I)

O_R

OFF

ON

Diagram of a molecular switch. The lambda viral DNA is integrated into the host-cell chromosome and termed a prophage (A); in the lysogenic state, the C_I gene is transcribed (wavy line) to make mRNA, and the C_I protein is made. This binds to the operator regions (O_R) and prevents cro mRNA synthesis. During lytic growth (B), as the virus replicates, the cro gene is transcribed and cro protein is made. This binds to the operator site (O_R) and prevents synthesis of C_I mRNA.

protein blocks cro synthesis in a reciprocal manner. One can readily see that this is a genetic or molecular switch: C_I turns off cro (and actually stimulates more C_I), while cro turns off C_I mRNA synthesis.

When lambda DNA enters a cell, there is a race between C_I and cro synthesis. If more C_I is made, that shuts off cro and shuts down all the lytic functions by inhibiting transcription at both major operator sites for the cro repressor protein of the virus. This favors lysogeny. If more cro is produced, that shuts down C_I synthesis. Now the O_L and one of the O_R sites and promoters are open for RNA polymerase to bind to, transcribing the genes of the virus and beginning a lytic infection. The genetic switch provides a functional shunt into the pathways of lytic or lysogenic infections.

Once the virus is in the lysogenic state, it has its 45,000 base pairs of DNA integrated into a specific spot in the three million base pairs of *E. coli* DNA, and only the C_I gene is expressed. The C_I repressor protein (there are about two hundred molecules of this protein per cell) binds to both major operator sites and keeps the lambda DNA from being expressed. No viruses are synthesized.

Several outside influences, however, can lead to the lytic pathway and lambda activation. Treatment of the bacterium with ultraviolet light, carcinogens, mutagens, or other agents that damage the DNA (break it or react chemically with it) induces the lysogenic bacterium to produce virus. Products of DNA damage and repair are able to convert an *E. coli* protein into an enzyme that specifically destroys the C_I lambda repressor protein (a protease). When, under these circumstances, C_I repressor is lost, the O_L and O_R sites now attract RNA polymerase at the promoter sequences, and cro (which shuts off more C_I synthesis) and the lytic functions are expressed. Virus is produced, and the cell lyses. In this way, a bacteriophage may live with its host for many generations, activating only in times of stress (DNA damage) to reproduce and infect healthy bacteria again.

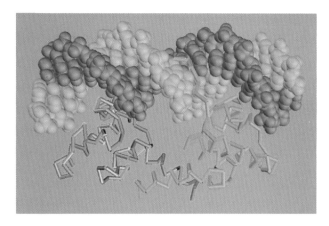

Computer-graphic representation of a portion of the lambda C_I protein recognizing the lambda DNA operator site (O_R). A portion of the lambda C_I repressor protein is shown as a dimer, with its polymer chain folded. When the protein touches the DNA (blue) and recognizes specific nucleotide sequences, it binds to that site and prevents RNA polymerase from transcribing the cro gene. The C_I protein also enhances mRNA synthesis from its own gene.

The lambda molecular switch that regulates the choice between two alternative pathways is an excellent model of a single cell that must eventually give rise to progeny that differ from each other (as in the development of a fertilized egg into various embryological tissues and, ultimately, into specialized cells—a muscle cell, say, or a brain cell). Genetic switches for tissue differentiation also use proteins that recognize DNA nucleotide sequences and turn genes and gene products either on or off. The ability to respond to outside cues and environmental changes by activating new sets of genes (which both animals and plants do all the time) also relies upon protein recognition of nucleotide signals in the genome of a host or virus. The study of these relatively simple viruses, the bacteriophages, and their host interactions therefore provided fundamental insights for molecular biology and virology.

وَيَعْتَرِقُونَ وَمِنْهُمْ مَنْ يَهْرُبُ مِنَ النُّورِ وَمِنْهُمْ مَنْ يَشْتِلَهُ

وَمِنْهُمْ مَنْ يَنْبَحُ مِثْلَ الْكِلَابِ وَبَعْضٌ مِنِ اقْتَرَبَ مِنْهُ نَصَبَهُ

تِلْكَ أَيْضًا وَقَدْ ذَكَرْنَا أَنَّهُمْ رَأَوْا إِنْسَانًا أَوْ إِنْسَانَيْنِ عَضًّا فَانْفَلَتْنَا

وَأَنْ أُورِدَ مُوسَى إِلَى هَذِهِ الْآفَةِ وَيَسُوسُ وَأَنَّ أَحَدَهُمَا عَضَّ وَالثَّانِي

بِهَذِهِ الْبَلِيَّةِ وَخَلَّصَ ع ع ع ع ع ع ع

وَأَمَّا الْأَخَرُ وَكَانَ السَّامِعُ خَالِصًا صَدِيقٌ لَهُ خَافَ الْآفَةَ فَلَا دَمَان

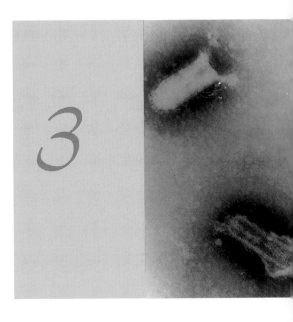

3

The Host Cell and Its Defenses

The way an idea eventually becomes an accepted truth is revealed by the stages through which it passes: First, it is said that "It can't be true"; then, "If true, it is not very important"; and finally, "We knew it all along."

Jonas Salk, 1958

From the turn of the twentieth century, when viruses were first discovered, it was clear that they could only be grown in their hosts. Tobacco mosaic virus replicated only in the cells of the tobacco leaf; yellow fever virus was obtained only from the lesions of infected people. While Koch's postulates could be satisfied for bacteria—most of which can be grown in pure culture, away from the host, just by feeding them glucose and salts on an agar surface—viruses are bound to their host organisms. Those who studied viruses needed to have a colony of bacteria, plants, or animals available to use as hosts. Some advances were made by adapting viruses from many different hosts to the embryonated chicken egg. A solution of virus could then be inoculated through a small hole in the shell; once the egg was resealed, an encapsulated source of chicken cells and a virus that had duplicated itself within them could be obtained. But adapting a virus to chicken cells often selects for rare mutants that differ from the original agent under study: what could flourish in the chicken cells, then, was often a virus changed from the original isolate inoculated into the egg. That was clearly not the way to satisfy Koch's postulates.

Left: This thirteenth-century Iraqi manuscript page, depicting a rabid dog biting a man, appears to be part of an early medical text. Written by Abdallah ibn al-Fadl of the Baghdad School, it states: "And they sweat. Some of them flee from the light and some of them take pleasure in it. Some of them bark like dogs. At times, one who approaches one of them may become similarly afflicted. People have reported that they have seen one or two persons who were bitten and escaped."

Above: EM of rabies virus. The mature virus particle has a bullet shape and a membrane—removed in one particle to display the helical symmetry of the inside.

Cells in Culture

The real breakthrough in technology occurred when it was appreciated that single cells obtained from embryonic tissues or even from adults could be grown outside of the host (that is, in culture) for several generations. So long as one supplied all the nutrients required for their growth, they would divide. Viruses added to such cells could attach, penetrate, uncoat, replicate, and reproduce. Best of all, many different species of cells—human, mouse, rat, monkey, and so forth—would grow, so most viruses could now be propagated in cell culture. It was even possible to obtain a pure culture of an animal virus to test against Koch's postulates. As we saw in Chapter 2, if a virus solution is diluted manyfold and just one or a few virus particles are added to a culture dish, each virus will attach to a cell, propagate itself, and be released to infect the adjacent cells. After several rounds of replication, the original virus focus will spread out to form a visible zone of dead cells. This is, of course, d'Hérelle's plaque. You will recall that a single plaque contains the progeny of a single parental virus: its linear descendants, or clones. It was now possible to isolate a virus from a lesion in an animal, plaque-purify the virus in cell culture, rein-

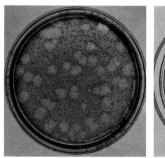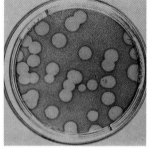

A   B

Virus-infected cells produce plaques. A: Chick embryo fibroblasts infected with Western equine encephalomyelitis virus (a togavirus). B: Human cells infected with poliovirus.

oculate the cloned progeny of the virus into an animal host, see if it caused a disease, and then reisolate the progeny virus. In this way, Koch's postulates were satisfied to show definitively that viruses do cause diseases.

Plaque assays also afforded a means to isolate mutant viruses in pure form. Some viruses use cell lysis for egress; let us suppose, for example, that a mutation arose in a population of these viruses such that the viral-encoded protein responsible for cell lysis became less efficient in breaking open cells and releasing viruses for additional rounds of replication. One might now expect two types of plaques. Efficient lysis of a cell would result in thorough infection of the neighboring cells and rapid spread of the virus: here the plaque would be large, encompassing many cells. But the mutant virus would not get out of the infected cell as efficiently and would spread to its neighbors slowly: this plaque would be small. Small-plaque viruses are well known, in fact, and by isolating such a plaque and placing it in solution, a clone of viruses descended from one mutant agent can be derived.

As cell-culture techniques were developed, several interesting properties of cells were noticed and recorded. Single-cell suspensions taken from a human embryo or newborn (placenta or foreskin) could be prepared and grown in culture on the surface of a glass or plastic dish. The cells replicated sixty to eighty times and then failed to divide any more. Human cells placed in culture for the first passage of growth are termed primary cell cultures; after passage and division, they are called secondary cultures. Such normal cells die at sixty to eighty generations and never form a permanent cell line. This experiment could be repeated in different laboratories, with cells from different people, in different preparations of nutrients. Apparently, human cells in culture have a defined and limited life span. Furthermore, the cells somehow remember the number of times they have previously divided. It is possible to take cells at twenty generations of growth and freeze them at minus 80°C in a suspended state. After several years, these cells can be thawed and grown in culture again. Cells that had

Does the limited number of divisions observed in normal human cells in culture reflect an inherent property of the aging process?

grammed to have a limited life span of sixty to eighty generations and that we use up these divisions with age. Aging would then be defined as a declining ability to reproduce our blood cells, skin cells, bones, and so forth. We may age at different rates because the program for such cell death, inherent in the DNA molecule, is different in different individuals; that is, longevity may have an inherited component. Not everyone agrees with this interpretation of these experiments, however; we do not yet know how the number of divisions a cell can undergo in culture relates to the normal functioning of the cell in a living animal.

Because these observations created a good deal of interest in the growth of normal cells in culture, some scientists examined in greater detail what happens when cells reach the sixty-to-eighty-generation barrier. It appears that the number of hours cells take to undergo division increases until the process ceases in the entire culture. The cells synthesize proteins, membranes, and adenosine triphosphate (ATP) and otherwise function, but they no longer divide.

Cells derived not from normal tissues but from a variety of cancerous tissues can also be grown in culture. In many cases, these cells fail to replicate extensively in culture. In a significant number of examples, however, the cancer cells in culture grow and divide forever (or at least, in the longest documented case, for forty years). These cancer cells do not enter a crisis period at sixty to eighty generations. They have lost the limited life span shown by normal cells in culture and instead can provide a permanent cell line—in effect, they produce immortalized cell lines.

As we shall see in future chapters, some viruses contribute to or cause cancers in animals. At this juncture, consider an interesting experiment designed to answer several questions: What would happen if we took normal human cells in culture, derived from the foreskin of a young boy, and infected them with a tumor virus—that is, a virus capable of initiating a cancer in an animal? Can a tumor virus produce an immortalized cell line? Can a tumor virus bypass the crisis period? When this

been through twenty generations will replicate forty to sixty more generations in culture; those that had been at fifty generations will replicate ten to thirty more generations in culture.

Perhaps more remarkable is the observation that cells taken from older humans (sixty to seventy years old) replicate for fewer generations in cell culture before death than do cells taken from newborn humans. Apparently this programmed count of the number of cell generations is affected by the age of the individual donating the cells. The count of the number of cell generations may be occurring in human beings as they live their lives. Some scientists believe that these experiments are measuring an inherent property of the aging process. They hypothesize that all cells are pro-

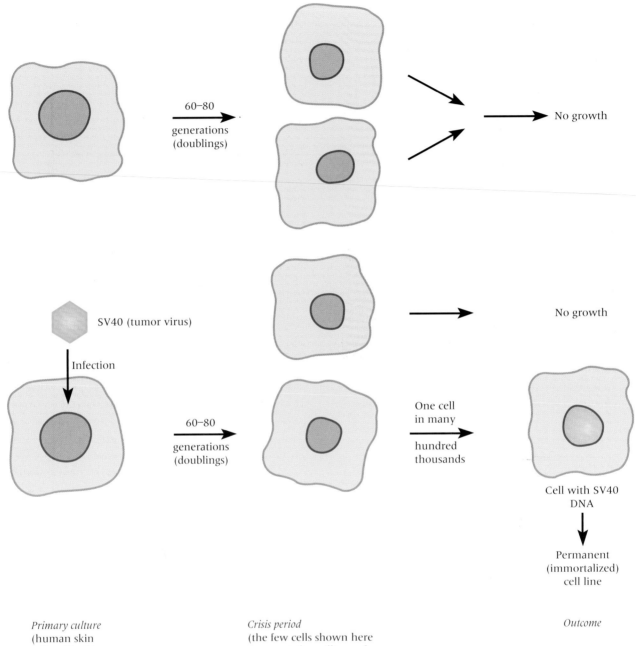

60–80
generations
(doublings)

No growth

SV40 (tumor virus)

Infection

No growth

60–80
generations
(doublings)

One cell
in many

hundred
thousands

Cell with SV40
DNA

Permanent
(immortalized)
cell line

Primary culture
(human skin
cells—fibroblasts—
from newborn)

Crisis period
(the few cells shown here
represent the millions of
progeny in the secondary culture)

Outcome

was done with a virus that can initiate tumors in rodents, simian virus 40 (SV40), the human cells in culture continued to divide for up to eighty generations and then went into the crisis period, slowing their division times. With normal, uninfected cells, this would end the experiment; they always fail to divide again. (This is the control for our experiment.)

But with the SV40-infected cells, a rare cell in the population—perhaps only one in a hundred thousand or a million—began to divide again in the culture dish, eventually taking over the entire cell culture. (Indeed, the only replicating cell in a population will inevitably be selected for in culture.) When the progeny of these rare cells were examined, they all contained a portion of the SV40 virus. The viral DNA was integrated into the chromosome of the host cell so that it formed a continuous DNA strand of human and SV40 nucleotide sequences. The viral DNA was, in effect, spliced into the human chromosome. Furthermore, this viral DNA encoded a viral mRNA that, when translated on the ribosomes of the host cell, synthesized a viral-encoded protein. As we shall see when we explore this virus in Chapter 5, this viral protein is indeed responsible for conferring upon this rare cell the ability to get through the crisis period and form a permanent cell line in culture. SV40 has the ability to immortalize cells in culture; they behave like some cancer cells derived from human beings.

This assay has become one of the ways in which we recognize tumor viruses in cell culture.

Schematic representation of cells in culture. If infected with a tumor virus like SV40, very rare cells will emerge from the secondary culture population to bypass the crisis period and form a cell line that can duplicate itself forever. These permanent lines contain—integrated into the chromosomes of the host cells—the viral DNA, which expresses the viral proteins required to confer immortality. Note that here and throughout the book, viruses are shown much larger in proportion to cells than they are in reality.

Developing such tests for tumor viruses eliminates the need for a host animal in research and permits the experimenter to study the process in greater detail. Although some researchers feel that the so-called immortalization test in culture measures a significant set of events that also occurs in cells as a cancer develops in the living host animal, others are skeptical. Definitive proof that the immortalization test measures an important property of cells, relevant to cancer, awaits more information.

Our Defenses Against Viruses

In the struggle of human beings, animals, and plants to survive and reproduce, the diseases caused by microbial parasites are a considerable burden. For complex animals and plants to thrive, they have had to devote a significant percentage of their genetic information and their energies to erecting defenses to neutralize or kill microorganisms. Human beings have developed physical barriers (skin, for example) and synthesize enzymes that attack and degrade bacteria—our tears, for instance, contain such an enzyme. We have specialized cells, the macrophages, that are attracted to bacteria, engulf them, and kill them (phagocytize them). Chief among our defenses, however, is the immune response to foreign organisms. Without the immune system, we could not live in the sea of microorganisms that surrounds us. It will be useful to describe a theoretical experiment that demonstrates the special elements and properties of the immune system.

We begin with four mice, which we number 1, 2, 3, and 4. Mouse 1 and mouse 2 are injected with a suspension of killed poliovirus; it will not cause any disease. Similarly, mouse 3 and mouse 4 are inoculated with a preparation of killed influenza virus, and these mice will also remain well. Two or three weeks later, mice 1 and 3 are injected (challenged) with live poliovirus, while mice 2 and 4 receive a live influenza virus challenge. Within a week, mouse 3 will be replicating large levels of

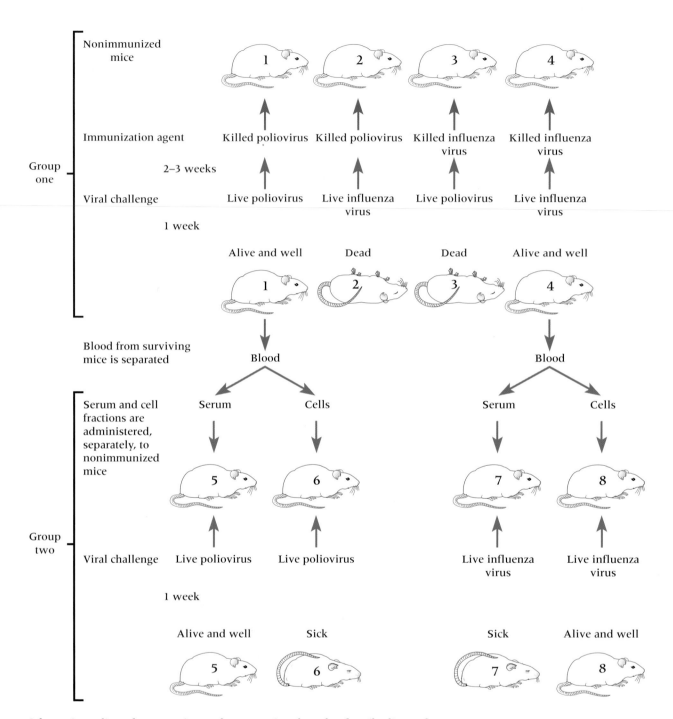

Group one

Nonimmunized mice

1　2　3　4

Immunization agent

Killed poliovirus　Killed poliovirus　Killed influenza virus　Killed influenza virus

2–3 weeks

Viral challenge

Live poliovirus　Live influenza virus　Live poliovirus　Live influenza virus

1 week

Alive and well　Dead　Dead　Alive and well

1　2　3　4

Blood from surviving mice is separated

Blood　Blood

Group two

Serum and cell fractions are administered, separately, to nonimmunized mice

Serum　Cells　Serum　Cells

5　6　7　8

Viral challenge

Live poliovirus　Live poliovirus　Live influenza virus　Live influenza virus

1 week

Alive and well　Sick　Sick　Alive and well

5　6　7　8

Schematic outline of an experiment demonstrating the role of antibodies and T-cell immunity in virus infections.

poliovirus and will become ill (actually, poliovirus will not attach to mouse cells or cause disease in a mouse, but this is a mental experiment, intended to be illustrative). Mice 1 and 4 will remain well, while mouse 2 will have severe flu symptoms.

This experiment demonstrates three important principles of the immune system: (1) the system is inducible—a prior exposure to a virus is required to provide good protection; (2) the system has specificity—influenza virus does not produce a response that protects against poliovirus, and vice versa; and (3) the system has memory—a prior exposure will provide protection for months or years.

We can next ask what it is that protected mouse 1 from poliovirus infection and mouse 4 from influenza virus infection. If we were to take blood samples from mouse 1 and mouse 4 and examine them under a microscope, we would note that each is composed of a clear fluid (serum), containing thousands of different proteins. Swimming in this fluid are a wide variety of cells. If we separate the serum from the cells, we can ask: Where is the activity that protects these mice from virus reinfection? A good way to begin doing this would be to inoculate the serum from mouse 1 into another mouse, number 5, and the cells from mouse 1 into mouse 6 (where mice 5 and 6 have never before been exposed to poliovirus). Now we can challenge mouse 5 and mouse 6 by inoculating live poliovirus into them. After a week, mouse 5 is well but mouse 6 is supporting some virus replication and shows some signs of disease. The first half of our experiment is complete. We conclude that the protective substance for poliovirus is predominantly in the serum. The protective agent, called antibody, is a serum protein. Had we infected mouse 5 or mouse 6 with influenza virus, both would have been ill. Antibody, as an aspect of the immune response, is specific in its mode of action.

We have just developed our first experimental test for antibody, so let us use it to make a few observations. First, if we examine a mouse that has never been exposed to killed poliovirus, the blood of this mouse will have no detectable antibody that

could protect the mouse from poliovirus. If we inoculate the mouse with killed poliovirus, antibody appears in the serum. Antibody is inducible in response to a foreign substance. The foreign substance, poliovirus, is termed an antigen. While mouse 1 was protected by antibodies made in response to an exposure to the killed poliovirus (mouse 1, we recall, was vaccinated), mouse 5 received only antibody, not killed virus. The antibody itself protects this mouse without a prior virus challenge. This is called passive immunity—that is, the passive transfer of antibody is to be contrasted with the active exposure to antigen (virus) that induced the synthesis of antibody. Passive immunity has no memory, and it decays with time. Injecting gamma globulin (a fraction of human serum with antibodies in it) provides passive immunity.

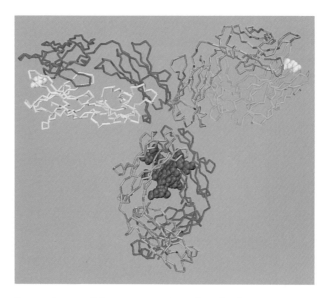

Computer-graphic representation of antibody structure. Antibody is composed of four protein chains—two identical light chains (here, yellow and green) and two identical heavy chains (here, blue and orange). The three-dimensional structure and folding pattern of these polymers are shown; the purple molecule is the attached carbohydrate. This antibody has two sites for combining with a virus (shown at right and left extremes where blue-yellow and orange-green chains interact).

We can next ask how the antibody protects against poliovirus infection. Here we could use the poliovirus plaque assay to watch poliovirus replicate in cell culture with or without antibodies present. When we mix poliovirus with serum from mouse 1 or mouse 3, no plaques are observed in the cell-culture dish containing mouse 1 antibody plus poliovirus, but lots of plaques appear on the cell-culture dish with influenza virus antibodies and poliovirus present. By recognizing the viral coat proteins, antibody combines directly with a virus particle, and the physical combination blocks one or more key events in the viral replicative cycle (attachment, penetration, uncoating, and so forth). Antibody can act in a cell-culture dish or in an animal to prevent the virus from reproducing itself.

We still have the other half of our experiment, however, to carry to its conclusion. The serum from mouse 4 (immunized and challenged with influ-

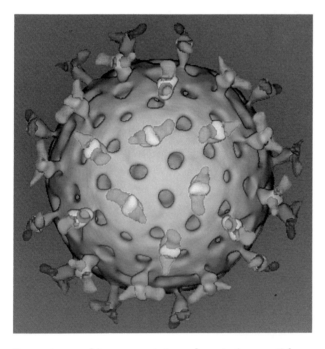

Computer-graphic representation of a rotavirus particle (blue) with antibody molecules (orange) attached to its surface, inactivating it. This shows the relative sizes of antibodies and viruses.

enza virus) or its blood cells are next inoculated into mice 7 and 8, respectively. A few days later, these mice are challenged with live influenza virus. Mouse 7, which received antibody alone, replicates some virus and is slightly ill (the antibodies work less well here), while mouse 8 is well and shows very little virus replication in its lungs and other tissues. In this case, the cellular fraction of the blood provided most of the protection against the influenza virus, while in the case of poliovirus, the antibody fraction in the serum provided the majority of the protection.

If we were to fractionate the blood cells into different components, we would find that small lymphocytes (white blood cells) called T cells had most of the ability to protect mouse 8 from the influenza virus infection. T cells get their name because they reside in an organ called the thymus for most of the later stages of their development. In the thymus, two major classes of T cells have been recognized and classified, based upon proteins that are found on their plasma membranes. These cell-surface proteins, called cellular determinants (CD), distinguish the two types of T cells—CD4 and CD8. They have different functions.

A CD8 T cell, often called a killer T cell, recognizes a foreign antigen on the surface of an infected cell and then kills that cell. CD8 T cells are the cells that protected mouse 8 from influenza virus infection. During its replicative cycle, the influenza virus synthesizes viral-encoded proteins, portions of which are displayed on the plasma membrane of infected cells in conjunction with cell-surface proteins specialized for this purpose. Throughout the infection, these surface portions of viral proteins are recognized by CD8 killer T cells, which target and kill the virus-infected cells, limiting the amount of virus synthesized. This reduces the level and spread of virus and terminates the infection rapidly.

CD4 cells, the other class of T cells, are often called helper T cells. They synthesize specialized protein factors called lymphokines, or interleukins, which are required for the optimal production of antibodies (produced by other white blood cells, called B cells) and for CD8 killer T-cell activity.

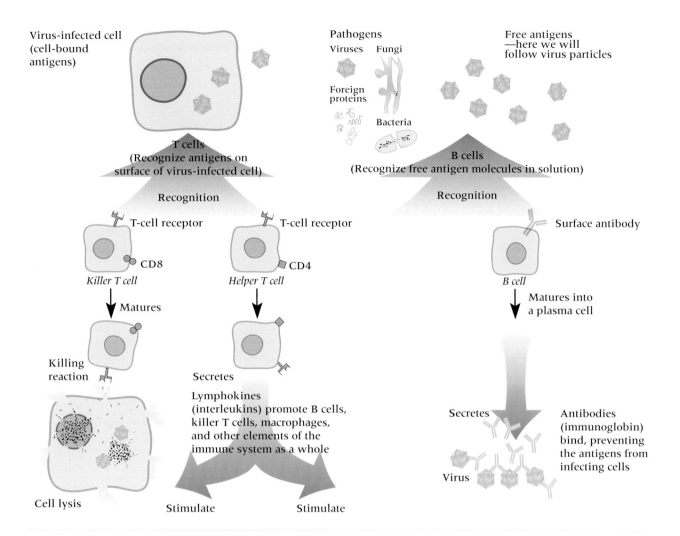

Virus-infected cell (cell-bound antigens)

Pathogens
Viruses Fungi
Foreign proteins
Bacteria

Free antigens —here we will follow virus particles

T cells (Recognize antigens on surface of virus-infected cell)

B cells (Recognize free antigen molecules in solution)

Recognition

Recognition

T-cell receptor
T-cell receptor
Surface antibody

CD8
CD4

Killer T cell
Helper T cell
B cell

Matures
Matures into a plasma cell

Killing reaction
Secretes

Lymphokines (interleukins) promote B cells, killer T cells, macrophages, and other elements of the immune system as a whole

Secretes

Antibodies (immunoglobin) bind, preventing the antigens from infecting cells

Virus

Cell lysis
Stimulate
Stimulate

CELLULAR IMMUNITY
(killer T cells and helper T-cell secretions)

Cellular immunity is a consequence of the activity of T cells. There are two types; both have T-cell receptors on their surfaces and recognize cell-bound molecules on the plasma membranes of infected cells. They neither respond to free antigens in body fluids nor secrete antibodies.

Killer T cell

Has CD8 receptor. The killer T cell takes infected cells (not free antigens) as its target. It binds to them and kills them, inactivating the pathogen: here, a virus. By boring holes and inducing cell lysis, one killer T cell can prevent virus replication in many host cells and slow virus spread in the organism.

Helper T cell

Has CD4 receptor. Response is the secretion of lymphokines, or interleukins (there is a great variety of these protein molecules). These feed back into the immune system overall by stimulating both B-cell and killer T-cell response: hence the helper T cell is a key factor for optimal function of both arms of the immune response.

HUMORAL IMMUNITY
(B-cell secretions)

Humoral ("fluid") immunity is a consequence of antibodies: serum proteins that are collectively called immunoglobin; they constitute about one-fifth of the proteins in the blood. B cells, recognizing free antigens in the body fluids, mature into plasma cells and secrete antibody molecules that bind to the antigens and inactivate them. Viruses, with many identical binding sites for antibodies (because of their structural symmetry), can form large aggregates when they react with immunoglobin.

The two arms of the immune response: cellular immunity and humoral (serum) immunity.

Thus, there are two arms to the immune response: (1) the B-cell production of antibodies, which are found in the serum and act by combining with antigens and neutralizing them; and (2) the generation of killer T cells (CD8), which recognize foreign proteins on the surface of virus-infected cells and destroy these cells. The CD4 helper T cells are required for the function of both arms of the immune system.

In general, antibodies protect us against many viruses as well as some bacteria, where they bind to the cell surface and help macrophages digest the bacteria. The CD8 killer T cells protect us against certain viruses, mediate tissue rejection of foreign transplants, and play a key role in combating fungal and some chronic bacterial infections. The two arms of the immune response work in a complementary fashion to protect the host. Individuals who have lost one or both arms of this response are immunocompromised; they are usually constantly battling the terrible effects of these parasites. Such a defect may arise from a genetic abnormality (a mutation in chromosomal DNA) or from a parasite attacking the immune system. The human immunodeficiency virus (HIV), which causes the acquired immune deficiency syndrome (AIDS), is such an agent, as we will see in Chapter 7; by replicating in and killing the CD4 helper T cells, it attacks both arms of our immune system.

protects us from additional occurrences of that infection. Viruses that specialize in infecting vertebrate animals, like humans, with well-developed immune systems face the problem of having limited access to the host; they can infect each of us only once. Inevitably, because of this limitation on reproductive capability, there is selection for viruses that can get around this problem. Some destroy the immune system of the host (HIV; Chapter 7); others have learned to hide in selected cells and recur periodically when the time is right (the herpesviruses; Chapter 4); and some have developed mechanisms to change their coat proteins, so that every exposure to the immune system is, in effect, a new event (influenza A virus; Chapter 8).

This constant evolutionary conflict between the host and its parasites, each seeking reproductive advantage, ensures that new infectious agents will arise continuously. From a clinical perspective, then, the goal is to find a way to expose individuals to a virus that doesn't cause disease and yet confers lifelong immunity. The first human disease for which this goal was achieved became, over three hundred years, an instance of the host's triumph. No struggle with disease could have a better ending than the story of smallpox and its recent complete eradication.

Vaccines

Over the centuries, medical observers commented on a curious correlation: once people had had a particular infectious disease, they rarely got it again. In many cases, prior exposure to an infectious agent results in lifelong protection. As we now know, these are the hallmarks of the immune system—inducibility, specificity, and memory. Obviously the best outcome was exposure to a virus that caused only mild distress and still resulted in lifelong immunity.

Our immune system functions in two ways: it helps us recover from a virus infection, and then it

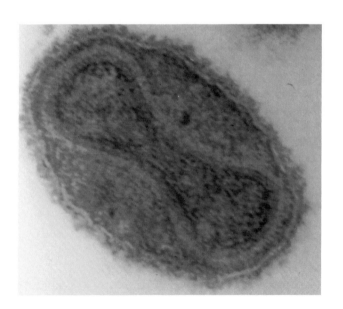

Smallpox Virus Vaccine

There are two types (strains) of smallpox virus, closely related by DNA sequences—variola major, which has a fatality rate of up to 25 percent, and variola minor, with a mortality rate of about 1 percent. The virus, which is transmitted by person-to-person contact, first replicates in the upper respiratory tract. A transient viremia (presence of virus in the blood) spreads the virus to the lungs, spleen, liver, and other internal organs; when virus titers rise in these organs, a second viral invasion of the bloodstream occurs about ten to twelve days after the first exposure. At this time (after the twelve-day latent period), the first symptoms appear: rash, fever, headache, malaise, and generalized aching. The virus now replicates in the epidermal cells of the skin. Vesicular eruptions follow in three to four days. All the symptoms are apparent in two weeks. While initially the virus is spread in respiratory secretions or by contaminated drinking or eating utensils, by two weeks the vesicles are full of infectious viruses. Eventually, the pustules rupture and form crusts; virus can be found for long periods of time in these healing lesions.

The recorded history of smallpox begins over two thousand years ago. The first reports originate in India and western Asia; the virus can then be traced to Japan and to Europe and North Africa at about A.D. 700. Its introduction to the Americas in the voyages of discovery and colonization was well documented by Spanish soldiers and priests. The terrible toll resulting from the infection of nonimmune Amerindians in the Caribbean (1507), Mexico (1520), Peru (1524), and Brazil (1555) had a great impact upon the histories of these peoples. Similarly, slave traders in the sixteenth and seventeenth centuries repeatedly introduced smallpox into Africa. Australian aborigines experienced an

Color woodcut from nineteenth-century Japan, showing the hero Yoritomo victorious over demon smallpox.

EM of smallpox virus. Among the largest and most complex of the viruses, it is brick-shaped and has an outer membrane and a dumbbell-shaped core; the viral DNA is wrapped about this core.

epidemic of smallpox in 1789, about a year after the arrival of the "First Fleet."

Methods of protection against this horrible disease date back to the tenth century. The Chinese powdered old crusts from the pustules and applied them by means of intranasal insufflation. Brahmins in India inoculated powdered dried crusts into the skin of persons with no disease. The Persians ingested crusts from patients; the Turks inoculated fluid from pocks into themselves. These practices were called inoculation, or variolation, and by and large they worked. For reasons that are not well understood even today, the disease produced by these practices was usually milder than naturally occurring smallpox, with case fatalities of only 1 or 2 percent instead of 20 to 30 percent. In 1718, the wife of the British ambassador to Turkey, Lady Mary Wortley Montague, introduced the practice of inoculation into England. During epidemics of mild disease (variola minor), crusts and vesicular fluids were taken from patients and used for variolation at later dates. The recognition of an attenuated disease and its use in vaccination was then widely employed in the British colonies.

In 1776, Edward Jenner made an interesting set of observations that led to a major advance. Jenner noticed that milkmaids in Gloucestershire, England, who acquired a mild disease from their cows called variolae vaccinae (cowpox), usually escaped smallpox even when this disease was epidemic in their community. By 1798, Jenner had inoculated people with cowpox and—by challenging them with an inoculation of smallpox—showed that these people were protected from that disease. The first vaccine, the principle of vaccination, and the use of a related animal (cow) virus for its attentuated effects in humans were all contributions that we owe to Jenner's insights.

The use of cowpox virus to immunize human beings caught on and spread throughout the world—although Jenner and his contemporaries could not, of course, be aware that what they were dealing with was a virus. The name given to the virus employed for smallpox vaccinations over the past fifty to one hundred years is vaccinia.

It had been assumed that this virus was derived from Jenner's cowpox virus and was a close relative. Recent studies, however, have shown that vaccinia viruses used in different parts of the world are very similar to each other but very different from the purported ancestor; some time ago, evidently, someone switched virus strains and the link to Jenner's cowpox virus was lost. All this happened before anyone had the tools to recognize differences between these viruses, nor was there any government regulation of vaccine viruses. We are

Edward Jenner, 1800, in a pastel portrait by J. R. Smith.

all very fortunate that the new vaccinia virus works to protect us from smallpox, although it is unrelated to cowpox.

In 1966, the World Health Organization (WHO) began a program of immunization to eradicate smallpox throughout the world. For this they had to develop a new strategy. It was simply not useful or possible to immunize entire populations. Even if one could introduce the vaccinia virus into every person in a country or area, some individuals experience poor virus replication and make few antibodies (or none) after a single exposure. Vaccinia was usually applied to a wounded area of skin to facilitate entry and replication of the virus, and the efficiency of application could vary. In any event, immunizing everyone in a population did

The Cow·Pock _ or _ the Wonderful Effects of the New Inoculation! _ Vide the Publications of ye Anti-Vaccine Society.

1802 cartoon, satirizing Jenner and "the Wonderful Effects of the New Inoculation!" On the wall of the room where cowpox vaccine is being administered hangs a picture of an Old Testament story: the worship of the Golden Calf.

not result in 100 percent protection from disease. Instead, WHO decided to identify individuals with disease (paying for this information, if necessary) and vaccinate all those in the immediate vicinity (the contacts). Teams of public health workers in India, Pakistan (which had epidemics as late as 1973), and all over the world identified cases and vaccinated all contacts surrounding each patient. At the end of World War II, most of Earth's inhabitants lived in areas endemic for smallpox; in Octo-

ber 1977, the last recorded case of a natural infection with smallpox was reported in Somalia. The reason to believe that smallpox is really eradicated is that humans are the only host for this virus; there are no known animal reservoirs. The endemic nature of smallpox could be kept in perpetuation only by passage from human to human—and that chain has been broken.

Based upon this logic, are further vaccinations for smallpox unnecessary? Vaccination is not with-

Photograph of Ali Maolin, October 26, 1977; he had the last naturally occurring case of smallpox (barring lab accidents) in the world.

longer vaccinate for smallpox. The population growing up today will be as susceptible as were the Amerindians of the sixteenth century. Curious differences in opinion creep into this set of decisions. In the United States, the general population is no longer immunized against smallpox, but recruits into the armed forces still are. Presumably, a virus can become a potent weapon.

Are there really no examples of smallpox virus anywhere in the world? Has this virus become extinct? Not exactly. In August 1978 (one year after the eradication of the virus in natural populations), two cases of smallpox were reported in Birmingham, England, the result of a laboratory accident where smallpox virus was being studied. This event made it clear that all stocks of smallpox virus needed to be either destroyed or made safe and secure. Most people agree there is a need for a reference standard of a smallpox virus, and one has been preserved under stringent controls at the U.S. Centers for Disease Control (CDC) in Atlanta, Georgia. It may be too early to be certain that the smallpox virus is extinct. If a persistent, unrecognized focus of this agent were to exist, the modern pace of travel and the increased changes in our environment could uncover an old virus and place it in a world of unimmunized people. The development of effective emergency measures and a constant vigilance will be necessary if we are not to be caught in a possible tragedy.

Modern Vaccine Strategies

At present, three different approaches are being taken to meet the goal of immunization against viral disease. Each has advantages and disadvantages. The first method, to which we will return, is the approach that conquered smallpox: immunization with living but weakened (attenuated) virus. The second method is to obtain large amounts of a virus and then kill it by exposure to a chemical (formalin) that prevents its replication but does not alter the virus structure or coat proteins. (As we

out its low level of complications: persons with immunodeficiencies suffer from progressive vaccinia disease, eczematous children sometimes contract a severe vaccinia virus infection, and postvaccinial encephalitis (a virus infection of the brain) is reported rarely but with a 30 percent mortality rate. Because there is some risk from the vaccine and apparently no risk of disease, many public health services worldwide have recommended that we no saw earlier in this chapter, it is the coat proteins of a virus that act as antigens to elicit an immune response from the host; antibodies produced by the host bind to the viral coat proteins and, in some cases, prevent adsorption to the susceptible host cell or, in other cases, permit a rapid elimination of the virus from the body. Similarly, killer T cells, recognizing a few virus-infected cells by pieces of foreign, viral-encoded protein on their cell surfaces, rapidly limit further infection by killing such cells.) Killed or inactivated virus can present proteins to the immune system—inducing an antibody response—without being able to replicate in the host and cause disease. Examples of killed intact virus vaccines are the Salk poliovirus vaccine, the influenza A virus vaccines, and rabies virus vaccine (see table below), which are inactivated by treatment with formalin.

In some cases, such as poliovirus, the antibodies that protect us are directed against a portion of the three-dimensional structure of the virus, the icosahedral particle. It is critical, therefore, to supply the virus particle as an antigen in order to elicit the correct protective antibodies. If poliovirus is broken down into its component parts and people are immunized with these coat proteins, antibodies are produced, but they fail to recognize the complex three-dimensional structure of the virus particle. Therefore, although they bind to the particle, such antibodies are not protective against a poliovirus infection. The live virus, then, must be treated with enough formalin to prevent it from replicating in cells, but not so much as to alter the structure of the virus.

Hepatitis B virus (see Chapter 9) contains a nucleoprotein core surrounded by a lipid mem-

Virus Vaccines Currently in Use

Virus	Disease	Type of vaccine
Poliovirus	Infantile paralysis (poliomyelitis)	Salk vaccine—inactivated virus particles Sabin vaccine—live, attenuated viruses
Influenza virus	Respiratory disease	Inactivated virus
Rabies virus	Rabies	Inactivated virus
Hepatitis B virus	Hepatitis; hepatocellular carcinoma	Subunit vaccine: surface antigen of the virus
Adenovirus types 4, 7	Respiratory disease; infectious pinkeye	Live, attenuated virus
Rubella virus	German measles (results in birth defects when pregnant women are infected)	Live, attenuated virus
Yellow fever virus	Yellow fever	Live, attenuated virus
Measles virus	Measles	Live, attenuated virus
Vaccinia	Smallpox	Live, attenuated virus
Mumps virus	Mumps	Live, attenuated virus

brane in which is inserted a viral lipoprotein called the S (surface) antigen. The S protein permits the virus to adsorb to the host cell—which, in this case, is a liver cell. Since immunity is generated by antibodies that recognize the S antigen of hepatitis B virus, it has been possible to immunize humans with just the S antigen in a lipid vesicle. The rest of the virus, with its nucleoprotein core and DNA, is not required for immunity and is not present in the vaccine. The S-antigen lipid vesicle is therefore a subunit vaccine (our third approach), in which only a single viral protein is employed.

It will be worthwhile, at this point, to return in greater detail to our first method of producing vaccines, employing live, attenuated viruses as the immunizing agents. A virus is said to be virulent when it causes pathology, or disease, in the host. Viruses can lose their virulence, producing only mild disease or no pathology; they are then termed attenuated. If a virus can be attenuated, then it can be used in a live form, which permits replication in the host, for immunization. Examples of this type of vaccine, besides smallpox vaccine, are the Sabin oral poliovirus vaccine (developed in the decade after the Salk vaccine); adenovirus vaccines (currently employed only for army recruits); rubella (German measles) virus vaccine; and measles, mumps, and yellow fever virus vaccines.

One of the most common ways to attenuate a virus is to introduce it into an unnatural host. (Recall that smallpox immunity was originally conferred by a cow virus.) The mumps virus, for example—a human virus—was inoculated into embryonated chicken eggs, where it grew poorly at first. After several passages (serial inoculations) in these eggs, rare mutant variants were selected for that grow better in chicken eggs. The mumps virus that emerged from this procedure grows well in chicken cells and poorly in human cells. It replicates well enough in humans to confer immunity, but not well enough to cause disease. Similarly, the yellow fever virus and measles virus vaccines were attenuated by replication in chicken embryo cells, while rubella virus and Sabin poliovirus vaccines were grown in monkey-kidney tissue-culture cells.

In all cases, mutant variants were selected that showed a reduced virulence in human beings. The example we will follow is poliovirus.

Poliovirus Vaccines

Poliovirus enters the host by the oral route. It may be found in contaminated food or water or in oral secretions. The virus passes into the intestine and replicates actively in the small intestine and associated lymph nodes. These nodes are the principal place of residence for B and T cells, so if the host has been previously immunized, a vigorous immune response at this stage limits the infection (the strategy employed by the Sabin vaccine). In the absence of such a response, the virus passes into the bloodstream and is spread throughout the body. Antibodies in the blood of immunized hosts (supplied by the Salk vaccine) are the second line of defense.

The virulent virus will replicate in a number of highly susceptible tissues of the body, including the motor neurons (cells that control movement) of the central and peripheral nervous systems. The virus kills these cells, and paralysis occurs. The latent period of this virus (the time from exposure to overt disease) is seven to fourteen days; the virus can be found in the bloodstream (viremia) by two to three days after exposure and in the central nervous system by six to seven days. Virus is excreted in the feces for several weeks, spreading to the sewer system (and, in rural or undeveloped areas, potentially back to the food and water supply).

The attenuated poliovirus used in the Sabin vaccine, by contrast, has a reduced neurovirulence; it replicates poorly in neuronal cells of the human spinal cord. The vaccine strain is administered to humans by placing the virus on a sugar cube (the oral route). The virus replicates well in the human intestinal cells, stimulating a local immunity and antibody production. The Sabin virus remains in the intestine for a long period of time and is excreted in the feces, so that the attenuated virus can be spread to household members of a newly immunized person. Because this polio vaccine is given to infants (first immunizations are at six to twelve weeks of age), adults are commonly exposed to the virus as they change diapers.

There have been rare reversions in the attenuated poliovirus, producing a more virulent form. This has resulted in cases of paralytic poliomyelitis in either the infant or the family member. In the United States, where oral poliovirus vaccine is used almost exclusively, there were 105 vaccine-associated poliomyelitis cases from 1973 to 1984—a remarkably low rate, considering that during this twelve-year period, 274 million doses of this vaccine were administered in the United States. Of these cases, thirty-five patients were the recipients of the vaccine; about half of these infants had an immune deficiency. The remainder of the known cases of poliovirus-related disease occurred in contacts of oral poliovirus recipients.

In the United States in the early 1950s, just before the Salk (killed) vaccine became available, about twenty-one thousand cases of poliomyelitis

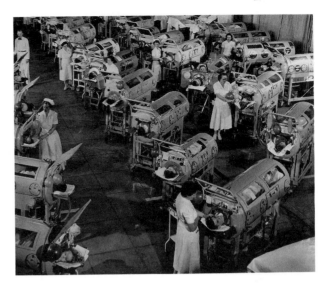

1952, the worst polio epidemic year in history: patients in iron lungs and rocking beds at a California medical center.

Jonas Salk administering his polio vaccine, 1954.

were reported annually, with peak incidence in the group five to nine years old. From 1970 to 1979, the average annual number of poliomyelitis cases was seventeen, and in the following decade, this average was fewer than ten. The incidence rate of poliomyelitis went from ten cases per year per hundred thousand people (in the 1940s) to 0.02 to 0.001 case per year per hundred thousand (in the 1980s). That is a remarkable record for the prevention of a disease.

Virus spread is a function of the hosts' behavior patterns; this is observed again and again in our recent past. With the introduction of blood transfusions, new sets of diseases and patterns of disease were observed. The life style of the host can result in the rapid spread of a virus (HIV and hepatitis B virus in the gay community are examples). This is also the case with poliovirus.

Before the twentieth century, paralytic poliovirus occurred in sporadic cases; epidemic poliomyelitis was unknown in China or other third-world nations. The virus was endemic all over the world, so almost everyone was infected and, therefore, immunized. Poor sewage systems and constant contamination of the water supply, along with the common use of human excrement (night soil) as fertilizer, ensured a constant exposure to this virus. New infants were protected by the mother's immunity, first in utero and then through breastfeeding. As the maternal (passive) immunity was lost with time, these infants continued to be exposed to poliovirus and acquired active immunity. Disease from this virus was rare, and when it occurred (before the twentieth century), it was against a background of high infant mortality.

It is paradoxical but not surprising, then, that the first epidemics or increased numbers of paralytic disease occurred in the Scandinavian countries and the United States, where excellent sanitation was put in place in the nineteenth century. In areas with high-quality water supplies and better alternatives for fertilizing crops, poliovirus did not have a chance to infect every infant. Children grew up without ever having been exposed to poliovirus—thus having no immunity. In the United States in the early 1950s, before the introduction of the Salk vaccine, the peak incidence group was children of elementary-school age, but two-thirds of the poliomyeletis deaths occurred in patients over fifteen years old. The susceptible population first encountered poliovirus long past infancy and obtained it from person-to-person contact, giving rise to an epidemic—as contrasted to the previous endemic pattern. It is ironic that better sanitation changed this disease from a rare, sporadic one to a curse that was feared by all who lived through the summer seasons of the 1930s, 1940s, and early 1950s. With modern sewer systems and an improved health standard came freedom from many diseases, but not without a price—a lost immunity.

Since the introduction of first the Salk (inactive) and then the Sabin (live, attenuated) vaccines worldwide, the incidence of disease has declined, but paralytic poliomyelitis has not been eradicated. This is partly because of incomplete vaccination of populations and, in several countries, poor public health systems to deliver vaccination. In some cases, rare changes (mutations) in the poliovirus circulating in a population can have an effect. There are really three distinct strains of poliovirus

Bas-relief dating from the eighteenth Egyptian dynasty (about 1500 B.C.). The shriveled leg of the priest (left) is a characteristic feature of a patient who has recovered from paralytic poliomyelitis.

(types 1, 2, and 3), closely related to each other but immunologically distinct—that is, immunity to type 1 does not protect us from type 2 or 3, and vice versa among the strains. We therefore must have a vaccine with all three poliovirus types. The type 3 poliovirus produces additional variants at a high rate. For example, in 1984 and 1985 in Finland, which had been free of paralytic poliovirus cases for twenty years, there were nine cases of paralytic poliomyelitis and one case of nonparalytic poliovirus disease. Virus isolated from healthy individuals and sewage showed that a poliovirus type 3 epidemic was in progress and that the virus was different from the type 3 poliovirus in the Salk inactivated vaccine used in Finland. Most people—but

Advantages and Disadvantages of Killed Versus Live, Attenuated Poliovirus Vaccines	
Killed vaccine (Salk)	**Live vaccine (Sabin)**
Produces humoral immunity (antibodies) in the bloodstream	Produces humoral immunity (antibodies) in the intestine and bloodstream
Does not induce intestinal immunity, so wild poliovirus continues to be transmitted by oral-fecal route	Vaccine virus spreads to household in feces; this is an advantage in spreading immunity, but sometimes the excreted virus is a mutant, more virulent form
Absence of living virus eliminates mutations to increased virulence	Vaccine virus may mutate toward neurovirulence
Failure to inactivate the virus produces the disease	
Absence of living virus permits use in immunodeficient hosts	Immunodeficient hosts can be hurt by this virus
Generally, repeated booster shots are required to maintain long-term immunity	Immunity is lifelong
Proven effective in large populations	Induces antibody very rapidly
	Oral administration is easy and efficient

not all—had enough antibodies to fight off this distinct but related epidemic virus. This unusual type 3 variant has not appeared anywhere else in the world; it was selected for by the particular immunized population. So long as a virulent, wild type of poliovirus exists and circulates in human beings, however, no nation can be secure. Our only defense is to maintain high levels of immunity through vaccination of a very high percentage of the world population.

This review of poliovirus vaccines prompts a curious observation. In the United States, the live, attenuated Sabin vaccine is employed exclusively, while Finland and a number of other countries have elected to use the killed Salk vaccine to immunize their populations. Why? The table above provides a list of the advantages and disadvantages of employing killed versus live, attenuated poliovirus vaccines. (The principles are the same for live or killed vaccine of any virus.) The key advantage of a

live, attenuated vaccine is that it replicates to high levels and induces long-term, even lifelong, immunity. Killed virus is rarely able to provide such long immunity, so booster shots are required. The key disadvantage of a live virus is that we have no control over the mutations and reversions to virulence that it may undergo.

As this chapter has demonstrated, the complex interaction between the host organism and its viral parasites is largely enacted as a drama that takes place at the level of the cell itself. As the electron microscope allowed us to visualize the players, so cell culture has enabled us to follow the progress of the action with greater precision. The next six chapters will each focus on the details of that drama as we understand it for six different groups of viruses, whose life cycles simultaneously illustrate the range of viral adaptivity and cause significant and distinctive pathology in their hosts—often, human beings.

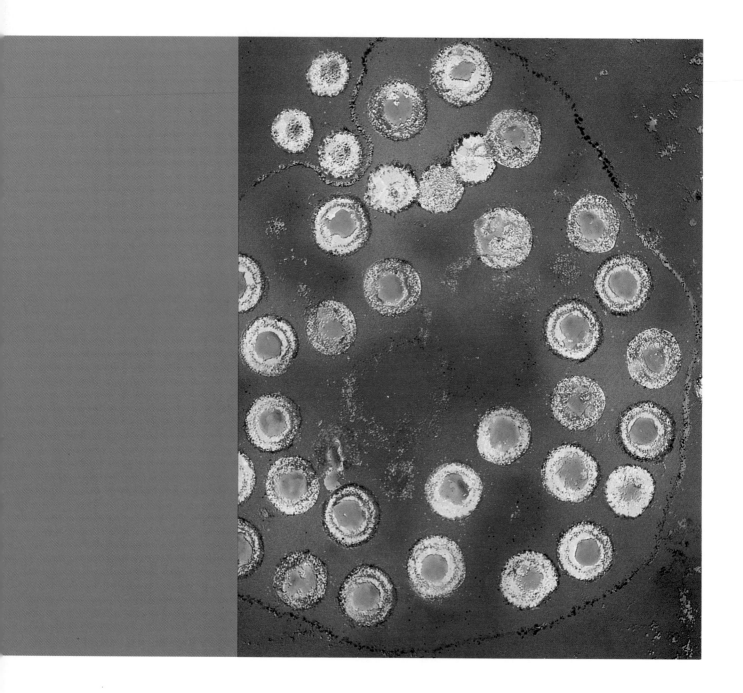

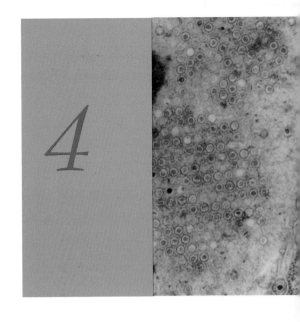

The Human Herpesviruses

O'er ladies' lips, who straight on kisses dream,
Which oft the angry Mab with blisters plagues,
Because their breaths with sweetmeats tainted are.

Shakespeare,
Romeo and Juliet, c. 1595

The word *herpes* comes from the Greek verb meaning to creep or crawl, and it aptly describes the spreading of skin lesions about the mouth that is characteristic of the disease later called herpes febralis or herpes excretins. In the last half of the nineteenth century, a herpes agent was definitively shown to transmit this disease, producing either facial or genital lesions. The medical literature of the mid-twentieth century expanded the list of pathologies associated with herpesviruses to include recurrent fever blisters (cold sores) of the mouth or genital region; keratoconjunctivitis (an infection of the eye); herpes infection of the newborn or the fetus in utero; disseminated (widespread) herpes infection in immunocompromised hosts; and encephalitis (infection of the central nervous system), which is often lethal. These syndromes are now recognized to be associated with herpes simplex viruses types 1 and 2 (HSV-1, HSV-2).

The most important biological capacity of the herpesviruses is their ability to hide within the body—in the case of HSV-1, within neurons (nerve cells)—in a quiescent state that can last for years. Because antibody cannot penetrate the plasma membrane of a cell, it does not neutralize latent HSV-1 in neurons. And because these infected neurons do not express HSV-1 glycoproteins on their surfaces,

Left: Herpesviruses appear here as yellow and green spheres within an infected cell, but the color is false. By definition, the electron microscope visualizes entities smaller than the color wavelengths of light. Every EM that employs color, therefore, does so arbitrarily.

Above: EM of densely packed intracellular particles of herpes simplex virus (HSV). The particles align symmetrically because of their icosahedral structure.

killer T cells cannot recognize that the nerve cells contain viral antigens and so do not attack these neurons. Neither arm of the immune system, then, protects us from a latent HSV-1 state. Herpesviruses live with us forever, despite our immunological defenses.

Herpes Simplex Viruses Types 1 and 2: The Nature of Latency

These two viruses are closely related, sharing a majority of their nucleotide sequences. The main difference between them is that HSV-1 has a predilection for oral and facial infections, while HSV-2 is most commonly found in genital lesions; but this distribution is not obligatory.

HSV-1 is transmitted from person to person in oral secretions or by direct contact, and it is commonly acquired early in life. The first infection is usually asymptomatic, although it occasionally affects the mouth and throat. Because it usually goes unnoticed, it has been difficult to determine when individuals were first exposed to this virus and how many of us have been exposed. A good way to test for a previous encounter with HSV-1, if a subclinical infection has occurred, is to look for the presence of antibodies directed against it in the blood. If antibodies in a serum sample inhibit HSV-1 growth and plaque formation in culture, we can conclude that there has been a prior exposure to the virus sometime in the life of the host. When this test was done with large numbers of samples (in New Orleans, Atlanta, and Houston), it showed that one-third of the population has been exposed to HSV-1 by age five. This prevalence increases to 70 to 80 percent of the total population of all ages.

There is a strong tendency for low socioeconomic groups to have higher rates of infection than middle and high socioeconomic groups; underdeveloped countries also have very high rates of childhood infection with HSV-1, resulting from poor sanitation and crowded living conditions. Although the symptoms produced by most of these early contacts are not apparent, rare complications can result in encephalitis or eye infections that can be serious or fatal. It is possible that some children who are mildly infected with HSV-1 will not develop antibodies, or will develop antibodies that are lost with time; they would, by this test, score negative for a previous exposure. For these reasons, antibody tests provide us with estimates of prior exposure to HSV-1, not absolute values.

When HSV-1 enters the oral cavity, it adsorbs to and replicates in the cells of the skin (epithelial cells) and in the mucous membranes of the lips and mouth. As the virus spreads, it may produce a local lesion or blister (vesicle) of infected cells, which eventually ruptures and heals with a crust, forming a cold sore. The virus in the lesion has been neutralized by antibodies, but other HSV-1 particles have escaped beyond the reach of the immune system.

Just under the skin in the oral cavity are clusters of sensory receptors used to discriminate between tastes or to feel pressure or heat. These receptors are connected, via long processes called axons, to the bodies of neurons that innervate our facial areas and provide sensory input. These neurons are clustered in structures called ganglia: way stations on the pathway that signals the central nervous system about the environment. It has been recognized for many years that HSV-1 resides in the trigeminal ganglia, in the region of the ear and temple. In 1905, Harvey Cushing, the founder of neurosurgery, was treating his patients for facial neuralgia—pain in the area innervated by these same sensory neurons—by the then-new approach of severing a branch of the trigeminal nerve. He noted that herpetic lesions consistently developed along the innervated regions of the sectioned branch of the axon, demonstrating a relationship between a virus and these ganglia.

We now know that HSV-1 can attach to receptors on the sensory nerve endings of the mouth that lead to the trigeminal ganglia. The viral nucleocapsid enters an axon and is then transported back into the body of the neuron and finally into its nucleus. The viral DNA remains in the neuron in an inert

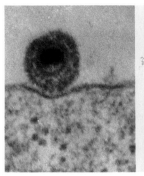
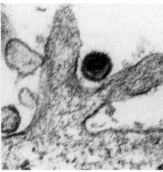

EMs showing a herpesvirus particle attaching to the plasma membrane of a cell. This step is mediated by the glycoprotein spikes on the surface of the virus. After fusion of the viral and cellular membranes, naked viral nucleocapsids move through the cytoplasm to the nucleus and enter it through the nuclear pores.

state (latency), with few if any of its genes being expressed. No mRNA from the genes of this virus has been detected in the ganglia, and if one isolates neurons from the trigeminal ganglia, breaks open these cells, and uses a plaque assay to test whether any infectious virus is present, the results are negative. Viral DNA is present in these cells, however, in a state of latency, isolated from the immune system of the host.

Even in the presence of cell-mediated or humoral immunity to HSV-1, therefore, reactivation of the virus from the latent state can occur, causing a new round of lesions. The induction of recurrent infections can be triggered by a number of external stimuli, including physical or emotional stress, fever (thus the name fever blisters), exposure to ultraviolet light in sunlight, tissue damage, and in some cases, immune suppression. Both genetic and environmental factors play significant roles in the frequency of recurrent herpesvirus activation, as do age and hormonal changes.

These external stimuli, somehow communicated to the infected neuron in the trigeminal ganglia, result in the activation of the viral DNA—that is, the expression of viral mRNA and the synthesis of viral proteins. Virus particles are produced, and the viral progeny are then transported back down the axon to a site at or near the original site of the skin infection (the portal of entry). Viral multiplication probably kills the neuron in the trigeminal ganglia; thus, each reactivation event uses up one or more cells in which HSV-1 is latent. Once in the epithelial cells of the skin, this virus will replicate, producing a fever blister. The reactivated viruses are the direct descendants of the original infecting virus, although years may pass between the primary infection and a reactivation. Indeed, the reactivated virus replicating in the epithelial cells does not even make a "round trip" back to the neuron; it is neutralized by the immune system and eliminated from the reinfection cycle. But the ganglia harbor other neurons containing HSV-1 DNA, poised to reactivate on another occasion.

Because infections with HSV-2 are usually acquired by sexual contact, antibodies directed against this virus are rarely found prior to the onset of sexual activity. Although HSV-1 can also be involved in genital infections, HSV-1 infections are usually less severe and less likely to recur. Their milder character indicates real biological distinctions between HSV-1 and HSV-2, despite the close relationship of the two viruses. The number of cases of HSV-2 infection in the United States has been difficult to assess accurately because doctors are not required to report this disease to the U.S. Public Health Service. Estimates range up to five hundred thousand new cases per year in the United States, and ten to forty million total cases. Even the lowest estimates imply a serious epidemic of HSV-2. Individuals with multiple sex partners are at high risk.

The disease follows the same general pattern of latency as HSV-1, but HSV-2 acquired in a primary infection replicates in the mucous membranes and epithelial cells of the genitalia instead of in the mouth and then travels through the axons of sensory neurons that innervate the genital region to the sacral ganglia at the base of the spine. Reactivation is triggered by hormonal changes, sexual activity, and other, less well defined, variables. With reactivation, the virus replicates again in the skin and mucosa of the genitalia, where it can be transmitted between individuals via sexual contact. Although such recurrences may result in painful lesions, they may be asymptomatic—a major factor in the spread of genital HSV-2 infection. A sensitive and specific antibody test for prior HSV-2 exposure reveals that 5 to 15 percent of the U.S. population has been infected with HSV-2; many adult patients are not aware of their latent infections. During the first year after primary exposure, recurrences can average 1.9 to 2.7 per 100 patient days, with more frequent recurrences in males. With time and age, these recurrences usually diminish and eventually stop.

One of the most troubling aspects of HSV-2 infections is that women may pass this virus on to a fetus or to offspring delivered through the birth canal. There are three recognized routes of infection of the newborn: in utero, during birth, or after birth. By far the most cases (75 to 80 percent) occur during birth, when the newborn is exposed to maternal genital secretions. Fifteen to twenty percent of neonatal infections appear to be acquired via contact after birth with the mother or another adult. Least frequently, a fetus is infected in utero when the virus replicates in the placental tissue. The consequences of newborn infection vary in severity and extent—encephalitis, skin lesions, keratoconjunctivitis, or widely disseminated infections. Because asymptomatic patients can be shedding virus, it has become common for women with a history of HSV-2 infections to deliver by caesarean section. By avoiding the birth canal, the infant escapes exposure to the virus.

HSV-1 and HSV-2 Structure and Replication

The linear, double-stranded chromosome of HSV-1 and HSV-2 is 152,000 nucleotide base pairs long: enough information to encode at least seventy proteins. In the virus particle, this DNA is wrapped about a protein core in the shape of a torus (dough-

A   B

A: Schematic diagram of a herpesvirus particle, showing the icosahedral inner core (with a view of the twofold symmetrical axis) surrounded by a membrane with glycoprotein spikes in it. B: EM of the icosahedral inner core of the virus with the outer membrane removed.

nut), which is surrounded by a protein capsid, 100 nm in diameter. The viral structural proteins account for about half the DNA coding capacity of the virus. The nucleocapsid, constructed from 162 capsomeres, forms an icosahedron that is first encased in a fibrous protein structure (the tegument) and is then surrounded by a lipid envelope from which several glycoproteins protrude. Some of these glycoproteins attach the virus to cell receptors and promote the fusion of the viral lipid envelope with the cell's plasma membrane. As the virion penetrates the cell, it is partially uncoated. The capsid is then transported to the nuclear pores, and the viral DNA is released into the nucleus.

A virion protein associated with the viral DNA next attracts a cellular RNA polymerase, which transcribes messenger RNA from a subset of five viral genes. From the mRNA, five proteins are manufactured in the cytoplasm of the infected cell, beginning a cascade of events. Some of these proteins (the immediate early, or alpha, proteins) promote the synthesis of additional viral gene products (delayed early, or beta, proteins); many of the latter are required for the replication of viral DNA. Both HSV types appear to have two kinds of genes that contribute to the duplication of their genetic mate-

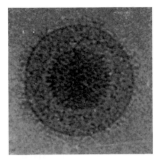

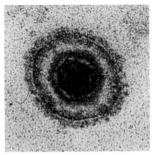

A   B

A: EM of an HSV particle prepared from frozen, hydrated samples; this preserves the spherical structure of the virion and permits visualization of the glycoprotein spikes on the surface. B: EM of cross section through an HSV particle, showing the nucleocapsid core, tegument, outer membrane, and glycoprotein spikes.

Computer-graphic representation of three views of the icosahedral core of a herpesvirus, reconstructed from EMs, showing the twofold, threefold, and fivefold rotational symmetries of the particle.

rial. One set produces proteins that polymerize DNA from A, T, G, and C; the other encodes a series of enzymes that synthesize those nucleotide precursors of DNA—thymidine kinase and ribonucleotide reductase, for example. These enzymes ensure a good supply of subunits for incorporation into the DNA polymer.

Thymidine kinase adds a phosphate group to thymidine (T), which is normally provided from the environment outside the cell, on the pathway to thymidine triphosphate, which is a direct precursor of DNA. Ribonucleotide reductase changes the sugar of the nucleotides from ribose (RNA precursors) to deoxyribose (DNA precursors). Curiously, these viral enzymes duplicate similar enzymes in the host cell, which has its own gene that encodes a thymidine kinase enzyme for converting thymidine to its phosphorylated form. This is also true for the cell's ribonucleotide reductase, which, again, functions like the viral enzyme. Because the cellular gene and the viral gene are evolving independently and act in very different environments—the uninfected cell versus the infected cell—the nucleotide sequences of the viral and cellular genes, while related, are distinct, even if they have the same func-

tion. Therefore, the amino-acid sequences of the viral and cellular thymidine kinases differ.

Why would the virus go to the trouble of obtaining and retaining a set of genes to encode enzymes that already exist in the host cell? Why not just use the host-cell thymidine kinase? We can devise an experiment to answer this question. Suppose we select for and obtain a mutant HSV-1 that has a defect in its gene for thymidine kinase and so fails to make this enzyme in an infected cell. If we grow this virus in tissue culture in cells that have their own thymidine kinase, the cellular enzyme substitutes very well: large amounts of the mutant HSV-1, unable to make its own thymidine kinase, are produced. In fact, one cannot tell this mutant virus from its parent that encoded the viral thymidine kinase gene. We might conclude from this experiment that the gene for thymidine kinase is not essential for the virus and that its presence is an evolutionary accident.

We should remember, however, that cells growing in a petri dish are quite an artificial environment for viruses. Viruses arise, evolve, and propagate themselves in the real world of cells in host animals—not in tissue-culture dishes. Replication in the neurons and skin cells of a human being must differ considerably from the relatively "good" life a virus has in a laboratory culture. If we introduce the mutant HSV-1 (without thymidine kinase) into an animal, it replicates very poorly and fails to establish a latent state, even though the host animal's cells contain thymidine kinase. A nonessential gene in one environment (cell culture) is indeed essential in another environment (the living animal). The lesson of this experiment is that, as Darwin first pointed out, changes in environment give rise to new forms of life. When the surroundings change, so too must a life form that had been optimally adapted for the previous conditions.

The reason that HSV-1 and HSV-2 have obtained their own, viral-encoded thymidine kinase is that these viruses must replicate in cells that are not dividing, such as neurons. Cells that are not replicating do not need to manufacture DNA, and therefore resting cells turn off some of the enzymes that produce DNA precursors. A virus entering such a cell has a very different agenda: it will be replicating its DNA within two to three hours, and it needs to have DNA precursors in great abundance. Viruses have solved this fundamental problem in two ways. Some, like the herpesviruses, carry their own genes for the enzymes that produce DNA precursors; other viruses have learned how to turn on the host-cell genes, forcing the cell to divide. Those viruses that employ the latter strategy are the cancer- or tumor-producing viruses, and we will review them in Chapter 5.

The second group of viral-encoded genes that contribute to the replication of HSV DNA encodes seven proteins that help to polymerize precursors into the DNA double helix. Chief among these is an enzyme termed DNA polymerase, which takes all four precursors—A, T, G, and C—and links them together in a continuous chain whose sequence is a copy of the parent-virus DNA. The cell has a DNA polymerase, too, but the virus prefers its own enzyme. In fact, the HSV-1 and HSV-2 virion proteins play a role in shutting off the synthesis of cellular mRNAs and proteins like the cellular DNA polymerase. By the time these viruses are replicating, the host cell is shut down, producing very few of its own products. The virus is monopolizing all the cell machinery (ribosomes, endoplasmic reticulum—ER—Golgi apparatus, and so forth) to produce viral proteins, not cellular ones.

High production of viral proteins really gets under way when the infected cell has synthesized thousands of copies of HSV DNA, many of which produce mRNA used on the cell's ribosomes to synthesize the third set of HSV proteins: the virion structural proteins (also termed the late, or gamma, proteins). In the first stage of assembly of the virus, the HSV-1 DNA is wrapped about its core proteins, and the 162 capsomeres condense around it to form an icosahedral shell in the infected cell nucleus. Viral glycoproteins are made in the rough ER, transported to the Golgi apparatus for the addition of more carbohydrate, and inserted into mem-

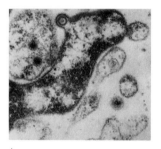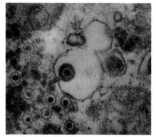

A   B

EMs of cells infected with HSV-1, showing maturing virions. A: Virus particle acquiring a membrane from the nucleus. Virus particles leave the nucleus via the pores and enter the space between the inner and outer nuclear membranes, which is contiguous with the endoplasmic reticulum (ER) of the cell. B: An enveloped capsid budding into the vesicles of the ER. Other particles in the cytoplasm have not yet acquired their envelopes.

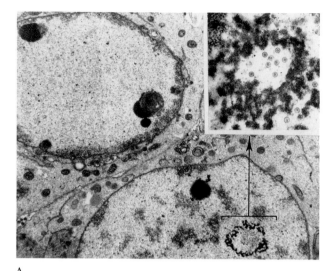

Chapter 4

A

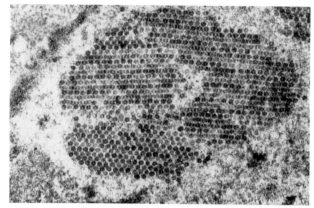

B

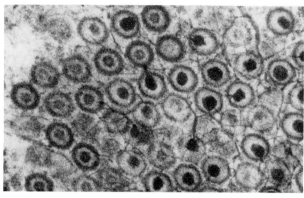

C

EMs of herpesvirus-infected cells showing various stages of viral assembly. A: In the nucleus, there appear to be sites for the assembly of the nucleocapsids. B: The density of progeny virus in a nucleus gets so high that paracrystalline arrays of virus particles form. C: These icosahedral particles in the nucleus are packaging their DNA. Some have a dense, dark center (DNA), and others have a lighter, less dense nucleoid; it is possible that this difference represents different stages of DNA condensation, but there is no evidence for this yet.

branes that will surround the icosahedron. Complete nucleocapsids leave the nucleus; they acquire one layer of lipid envelope in this process and another—with its glycoproteins—during passage out of the cell. The assembly of viruses continues for hours until the cell falls apart in exhaustion.

In the human body, virus infections are ordinarily contained by the immune system. In normal individuals, as we have noted, recurrent infections begin by passage of HSV down a nerve axon into skin cells. All this time the virus is isolated from the immune system, but once a lesion recurs, the virus freed from the dying cells stimulates a vigorous immune response: antibody and killer T-cell levels increase after each recurrence. This strengthened immune response clearly helps to keep the shed virus from reinfecting adjacent cells of the lip or

genitalia. Individuals who are immunosuppressed, however, are at risk of having the virus reactivate and spread throughout the body.

Until the last five years, there was very little treatment of HSV infections available. This changed dramatically with the development of acyclovir, a drug that can effectively inhibit three herpesviruses: HSV-1, HSV-2, and herpes varicella-zoster virus, the cause of chicken pox. The drug's crucial attribute is its selective toxicity: it kills the virus but not the host organism, by taking advantage of subtle differences in the viral and cellular versions of two key enzymes.

The acyclovir molecule structurally resembles deoxyguanosine, the G subunit in DNA. The HSV-encoded thymidine kinase mistakes acyclovir for guanosine and adds a phosphate group to the molecule. The viral enzyme phosphorylates acyclovir about two hundred times faster than the host-cell-encoded enzyme can, so an infected cell soon contains several-hundredfold more acyclovir phosphate than does an uninfected cell. This analog of the normal G nucleotide is phosphorylated twice more by cellular enzymes to produce acyclovir triphosphate, which now competes with guanosine triphosphate for incorporation into DNA. Here again, a viral enzyme—DNA polymerase—prefers the acyclovir triphosphate to the normal monomer, guanosine triphosphate, and incorporates the analog into its DNA. But once in a DNA polymer, acyclovir stops the growth of the chain—the HSV-encoded DNA polymerase binds very tightly to the acyclovir analog at the end of a DNA strand, becoming irreversibly inactivated. The host cell dies, but less virus is produced, and the immune system stops the spread of fewer viruses more effectively, allowing the organism as a whole to recover.

As of 1991, acyclovir has been used for as long as five years by people who have had recurrent HSV-2 infections. Taken orally two to four times a day, it has eliminated or reduced recurrences. It has shortened the healing time for lesions and reduced the level of virus shedding. Moreover, the drug has shown minimal toxic effects over the five-year period.

The only cloud in this bright sky is the appearance of HSV mutants that are resistant to the drug. Such mutants have been observed when the virus is grown in cell culture in the presence of acyclovir and—more rarely—in viruses isolated from human hosts taking acyclovir. Alternatively, a gene mutation may arise that eliminates *all* viral thymidine kinase activity. These mutants may well have diminished ability to replicate in host animals. Acyclovir mutations may also arise, especially in immunodeficient hosts, in the viral gene for DNA polymerase, making it more resistant to the drug so that the mutated virus replicates much less poorly in the presence of acyclovir than does normal HSV.

In practice, however, there has not been a high rate of drug-resistant HSV mutants in the population. The largest group of people to take acyclovir—individuals on long-term prophylaxis to prevent

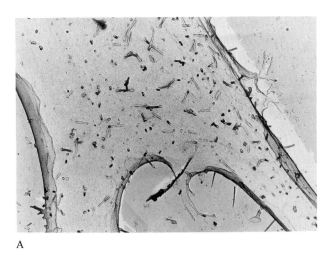

A

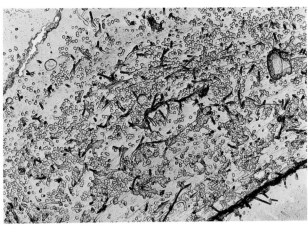

B

A: EM showing a section through the surface of an uninfected cell's plasma membrane, with normal projections and vesicles. B: EM of a section of a herpesvirus-infected cell shows a surface with thousands of virus particles leaving to infect other cells. This EM was obtained ten hours after the start of the infection.

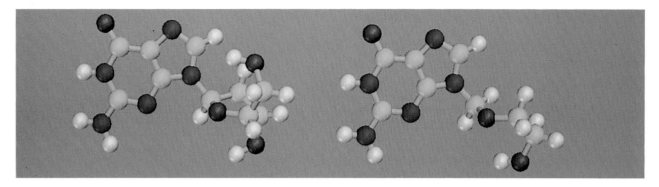

Computer-graphic representations showing the similar chemical structures of guanosine (left), a normal component of DNA, and the drug acyclovir (right).

frequent recurrent HSV-2 genital herpes—have normal immune systems that keep the virus levels in the sacral ganglia very low; the chance of random mutations occurring is proportionately low. Drug-resistant virus results more commonly from acyclovir treatment of immunosuppressed patients with severe HSV infection and proportionately high virus levels. For these patients, there is a great need for additional drugs of this type. (Only five antiviral drugs have been approved in the United States for systematic use, some of which will be discussed in later chapters.) By and large, however, acyclovir has been the single best chemotherapeutic agent made available to date against *any* viral pathogen.

Varicella-Zoster Virus

The third human herpesvirus causes two different diseases, chicken pox and shingles. For many years the connection between the two went unrecognized, and they were thought to be transmitted by two distinct viruses, whose separate names have now been linked. *Varicella,* a diminutive of variola (smallpox), reflects the rash and vesicles common to both smallpox and chicken pox, but there is no relationship between these two viruses or diseases. The term *chicken pox* is said to be from the French *chiche* (chickpea), describing the size of each pock, but this is speculation. *Zoster,* which was thought to be an independent disease-causing agent of shingles, got its name from the Greek word for girdle, describing the pattern of the rash typically formed across a patient's back.

Chicken pox, usually a childhood disease, is endemic all over the world. The virus replicates in the upper respiratory tract, after entering the body through the oral cavity. After a few days, the virus breaks out of the tissue into a viremia and replicates to high concentrations in organs throughout the body. A secondary viremia follows, seeding the virus into the skin and resulting in the characteristic rash. At this point, virus is present in the pocks and the patient is very contagious. Three days after the appearance of the rash, the immune system—employing both killer T cells and antibody—clears the virus, and the pocks heal in the next weeks.

A single primary infection is usually sufficient to protect us from ever having chicken pox again. But varicella-zoster is a herpesvirus, with the capacity for latency. During replication in skin cells, the virus enters the sensory neurons leading to the vertebral ganglia, which innervate the sides and back of the trunk. In most people who have had chicken pox, the virus will be dormant for life.

In some individuals, however, the virus reactivates and travels down the sensory neuron axons to the skin, where it replicates and forms the painful rash characteristic of shingles. The girdle pattern emerges as the virus follows the tracks of the sensory nerves. Several conditions appear to stimulate reactivation of this virus, including Hodgkin's disease and other lymphomas, immunosuppressive drugs, spinal cord trauma, stress, and heavy-metal poisoning. Shingles also tends to occur in older, rather than younger, people. Reactivation may occur once or a few times over a lifetime. Since varicella-zoster herpesvirus encodes the information for its own thymidine kinase and DNA polymerase, acyclovir treatment has been used to resolve shingles recurrences and ease the pain.

Initially, the relationship between the virus for chicken pox and the one for shingles was not clear. But in 1932, two scientists (in Germany and in Great Britain) independently inoculated filtered extracts from the eruptions of zoster patients into children and produced chicken pox. That experiment is not even good proof that the same virus causes both diseases—Koch's postulates would require us to isolate the virus in pure form and reproduce the disease from that isolate. (Today, the same virus has indeed been isolated in pure form from patients with either shingles or chicken pox.) To read of such experiments in the scientific literature of sixty years ago, however, serves as a reminder of how much we have changed in our ethical considerations with respect to the patient.

The Epstein-Barr Virus

Epstein-Barr virus (EBV), the fourth human herpesvirus, was first described by M. A. Epstein, B. G. Achung, and Y. M. Barr in 1964 as part of one of those wonderful scientific detective stories that take many twists and turns. The thread that led to the discovery of this virus was first picked up in the field of epidemiology, which attempts to correlate a subset of variables with the distribution and determinants of disease. The founder of Western medicine, Hippocrates, recognized and described this discipline in the fifth century B.C.:

> To investigate the origins of disease, one should first study the effects of the seasons, of the hot or cold winds, of the soil, naked or wooded, of the rains. Secondly, one will consider most attentively the way the inhabitants live, drink, eat, are indolent or active in exercise or labor.

De Contagione, published by Fracastorius of Verona in 1546, laid the foundations of modern epidemiology by investigating the distribution of events affecting the health of human populations. Fracastorius observed that epidemics of plague and syphilis were spread by seeds (*seminaria*) transmitted from one person to another by specific behaviors, and he recommended avoidance of exposure. These ideas, unpopular at the time, were as close as sixteenth-century scientists got to a germ theory of disease, and they formed the bases for valid epidemiological associations.

Working in this tradition, Denis Burkitt, an English surgeon from Makerere University College Medical School and Mulago Hospital in Kampala, Uganda, was studying the pattern of certain cancers in Africa during the 1950s. In particular, he noted that several tumors found in children and previously seen as unconnected should properly be classified as a single malignant lymphoma—a cancer of one of the white blood cells; in this case, the B cell that normally synthesizes antibody. Once this tumor syndrome, subsequently called Burkitt's lymphoma, was recognized, certain epidemiological facts became apparent. Nearly all cases occurred in children between two and fourteen years of age, with peak incidence at age five. The disease was found predominantly in Africa, although a few cases were known in New Guinea, and it appeared to have a geographical limitation. All races and tribes in Africa were susceptible, and even though the tumor did not exist in India, it occurred in Indian children who lived in Africa. Burkitt's lym-

Dr. Denis Burkitt (seated) and M. A. Epstein.

ation of Nyasaland and Southern and Northern Rhodesia [now Zimbabwe], the syndrome is found only in or near the great river valleys, and on the shores of Lake Nyasa. (4) It is common throughout the coastal plain of Mozambique. (5) It is virtually unknown in South Africa. (6) This pattern of distribution does not coincide with population densities.

The factors associated with cancer incidence were high temperature (low altitude), heavy rainfall, and abundant sources of water. A map showing the cases of these tumors followed the African malarial belt. Malaria is a disease caused by a mosquito-transmitted parasite, and areas of Africa where malarial eradication campaigns have reduced the mosquito population, such as the offshore islands of Zanzibar and Pemba, do not have high rates of Burkitt's lymphoma. Burkitt, appreciating this, stated, "The fact that the tumour distribution is dependent on climatic factors strongly suggests that some vector, perhaps a mosquito, is responsible for its transmission. This would naturally suggest that some virus may be the responsible agent."

Burkitt suspected a viral agent because a large group of viruses—the arboviruses, now called togaviruses—were known to be transmitted to humans by mosquitoes. One example is the yellow fever virus, the first virus of humans shown to be a filterable agent that causes disease. Although the association of a virus with mosquito-borne transmission was natural for an epidemiologist, the notion that a mosquito-borne virus could cause or contribute to a cancer was a surprising and original suggestion. It caught the attention of virologists.

The idea that some cancers could be caused by certain viruses was not a new one to the community of virologists. V. Ellerman and O. Bang, working in Denmark in 1908, demonstrated that they could transmit chicken leukemia from one animal to another by injecting an extract of leukemia cells, filtered to remove bacteria. Their work did not receive the attention or credit it deserved, however, because at that time leukemias (cancers of the

phoma represented more than 50 percent of the cancers observed in Ugandan children.

These first reports, in 1958, had little impact on the community of virologists. But in 1962 Burkitt, continuing his epidemiological studies, made several additional observations based upon a survey of tumor incidence as a function of location in Africa:

(1) Throughout Uganda, Kenya, and Tanganyika the tumour can occur anywhere except at altitudes of more than about 5000 feet, with the possible exception of the southern part of Tanganyika. (2) The offshore islands of Zanzibar and Pemba are a notable and significant exception. No case of this syndrome has been observed in these islands with a population of more than a quarter of a million. (3) Throughout the Feder-

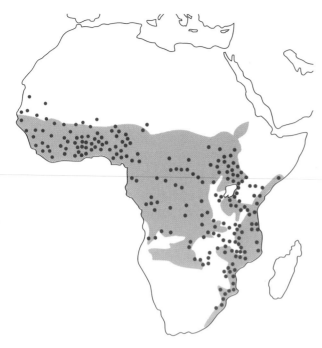

Burkitt's map of Africa, showing the geographical locations of the cases of lymphoma he analyzed and published in 1962 (dots). The cases fall within the malarial belt (shaded area).

blood cells) were not designated as types of cancer. This story, of course, illustrates the compromised impact of a discovery made at the wrong time.

In 1910, however, a farmer from upstate New York came to visit Peyton Rous at the Rockefeller Institute and showed him a Plymouth Rock hen with a large solid tumor of the connective tissue (a sarcoma or fibrosarcoma). Rous found that injecting some of the tumor cells into a healthy Plymouth Rock chicken induced a new tumor of the same type. Furthermore, he showed that it was possible to transmit this cancer by a filtered—cell-free, bacteria-free—solution made from tumor extracts. The Rous sarcoma virus, as this agent came to be named, was the first tumor virus identified, and it was to play a central role in the development of our current understanding of cancer.

By 1964, when Burkitt suggested that a virus could cause a human cancer, a large number of vi-

ruses were known that cause cancer in animals. But no human tumor virus had ever been found or isolated, so humans remained the exception to a growing list of hosts that could acquire cancer from a virus. Furthermore, there was another problem with Burkitt's hypothesis. Disease-causing viruses transmitted to humans via mosquitoes were all classified as togaviruses, and although several different groups of animal viruses—such as the papillomaviruses and the retroviruses—could be shown to cause cancer in their hosts, no togavirus had ever been associated with cancer in animals.

These objections were not allowed to stand in the way of the isolation of a human cancer-causing virus. In 1964 Epstein and Barr—and, independently, R. J. B. Pulvertaft—reported that they could take the cancerous B cells (lymphoblastoid cells) from a Burkitt's-lymphoma patient and grow them in cell culture for an unlimited length of time. These B cells from Burkitt's lymphoma, originally derived from patients in Africa, provided a permanent cell line that could be studied in the laboratory. The initial search for a virus in these cells turned up sporadic isolates of vaccinia virus (from vaccinations) or other contaminating agents; there was no consistent pattern until, fairly reproducibly in culture after culture, the same type of virus particle was seen in the electron microscope and classified as a herpesvirus. Epstein, Achung, and Barr reported their results in 1964. Once again, the conclusion reached did not really fit the notions of the times, for a herpesvirus had not been associated with animal cancers in the past. Still, several questions remained. Was this herpesvirus HSV-1, HSV-2, or varicella-zoster virus? Was it a new herpesvirus? What was its relationship to the tumor?

To answer these questions, Werner and Gertrude Henle, a husband-and-wife team of virologists working at the Children's Hospital of Philadelphia, developed a test to find antibodies in patients with Burkitt's lymphoma. They showed in 1966 that most—probably all—patients with this lymphoma had antibodies that recognized an antigen in the B-lymphoblastoid cells grown in culture. These antibodies did not react with normal B cells, only with cells derived from Burkitt's lymphoma.

Gertrude and Werner Henle.

Here was a potential test for the agent in Burkitt's-lymphoma cells, which had come to be called the Epstein-Barr virus.

Then came a rapid series of surprises for the Henles. First of all, most Africans, whether they had Burkitt's lymphoma or not, had antibodies that detected antigens in Burkitt's-lymphoma cells in culture. The difference between the antibodies from patients with lymphoma and those from people without lymphoma was that the level of antibody in those with the disease was at least tenfold higher than in those without the disease. The next unexpected observation was that most people in the Henle laboratory had antibodies that detected antigens in Burkitt's-lymphoma cells, although some people were clearly negative for such antibodies. The presence of antibody could not be ascribed to a hypothesis that everyone had caught the virus from the cell cultures in the laboratory, because individuals who had never been in the laboratory were also found to have antibodies directed against Burkitt's-lymphoma cells. In fact, random samplings from around the world showed a high percentage of people with antibodies that detected antigens in Burkitt's-lymphoma cells. The levels of antibody in average populations were low, more like the normal African population than like the lymphoma patients.

Virologists were initially convinced that these antibodies were directed against a common virus like HSV-1, which had somehow become a passenger in the tumor cells. But tests clearly showed that the antibodies that identified a herpesvirus antigen in Burkitt's-lymphoma cells did not bind to HSV-1, HSV-2, or varicella-zoster antigens in virus-infected cells. The Epstein-Barr virus was a new, presumably human, herpesvirus and was evidently a common agent throughout the world. A subset of people from specific geographic locations who had very high antibody levels to this virus also had a B-cell lymphoma.

There are times in research laboratories when many new results are found and a true puzzle emerges that you feel is really important. The clues are in, but the puzzle doesn't yet make sense. You need to ask the right next question, design the right experiment. It is often at these critical times that something disastrous happens to the progress of laboratory work. In this case, the Henles' technician, who maintained the cells in culture, contracted infectious mononucleosis and was out for several months. When she returned, a curious observation was made. The Henles, like all good virologists, often took blood samples from everyone involved in their research, so that they could monitor foreign viruses brought into the laboratory that might contaminate experiments and, conversely, so they could control for possible infection of the workers by the viruses and cells under study. The technician who had recovered from infectious mononucleosis acquired very high levels of antibodies directed against the Burkitt's-lymphoma B-lymphoblastoid cells, though she was completely negative for such antibodies before her illness. Furthermore, when the technician's B cells were cultured in petri dishes, they grew forever in cell culture—they were immortalized. These B cells contained an antigen that reacted with antibodies from Burkitt's-lymphoma patients. Finally, the B cells in culture derived from this technician secreted virus particles that were identical to EBV from Burkitt's lymphoma. These observations suggested to the Henles that EBV could be the cause of infectious mononucleosis.

This postulate was first proven to be the case in a seroepidemiological study carried out at Yale University. At Yale, blood samples were taken from a large number of entering freshman students. Four years later, medical histories and additional blood samples were taken. The samples were then tested to see if they included antibodies that combined with Burkitt's-lymphoma cells (from Africa) or with the B cells from the technician in the Henle laboratory. An interesting correlation emerged. Some students had antibodies directed against the EBV antigens in Burkitt's-lymphoma cells prior to admission at Yale, and some did not have such antibodies. At graduation, some of the students who had not had these antibodies as freshmen now showed antibodies directed against the EBV antigens; a large percentage of these students also reported having infectious mononucleosis sometime during their four years at college. The correlation between the acquisition of antibody and the clinical symptoms was a good one, and no one who had antibody to EBV at the time of admission had experienced the disease during the four college years. This and other seroepidemiological studies are excellent evidence—since we cannot use Koch's postulates with humans—that EBV causes infectious mononucleosis.

Exposure to EBV at an early age, as happens often in third-world countries and in lower socioeconomic groups, does not appear to cause a clear disease. EBV infection early in life is most often asymptomatic. For people in higher socioeconomic groups, the virus is usually first acquired in the teenage or college years. Transmitted in saliva by oral contact, infectious mononucleosis is sometimes called the kissing disease. EBV replicates actively in the epithelial cells of the mouth, tongue, salivary glands, and entire oral cavity. A person may have few symptoms at this time and yet be very infectious.

Spreading to the lymph nodes that surround the oral cavity, where the B and T cells of the immune system reside, the virus can infect B cells (but not T cells, which do not have the proper receptors for EBV adsorption). Once inside such a B cell, the virus expresses a subset of its proteins, both in the nucleus (these proteins are termed Epstein-Barr virus nuclear antigens, or EBNA) and in the cellular plasma membrane (these are the latent membrane proteins, or LMP). The expression of these viral proteins stimulates a vast number of B cells in these lymph nodes to replicate, producing identical copies of themselves (clones). Because many different B cells are infected, this response is called polyclonal B-cell growth—a phase of the disease that may begin a long time (the latent period) after the initial exposure to EBV and that results in a sore throat, swollen lymph nodes, fever, aches, and pains. In many cases, virus replication and disease occur in the liver and in the spleen (sites of B- and T-cell storage). The swollen lymph nodes and infiltration of mononuclear B cells in the nervous system create extreme malaise: a "ready to quit" feeling that can persist for a surprisingly long time.

In the great majority of cases, however, the immune system comes to the rescue. The epithelial cells and the polyclonal B cells—all expressing the viral-encoded LMP glycoprotein in their plasma membranes—are recognized by the killer T cells, and an attack is mounted to kill EBV-infected cells. During this time, with T cells killing B cells, the immune response of a patient is quite abnormal, producing atypical T cells and antibodies that confirm the diagnosis of infectious mononucleosis. The T cells slowly reduce the number of infected epithelial and B cells, and virus replication in the oral cavity comes under the control of the immune system. Antibodies directed against the EBV antigens are made, and the infection is resolved.

Nowhere is the value of the immune response as clearly demonstrated as in the case of infectious mononucleosis. Individuals who are immunodeficient, either for genetic reasons or because of AIDS or chemotherapy, often develop fatal B-cell lymphomas after EBV infections. The virus enters the B

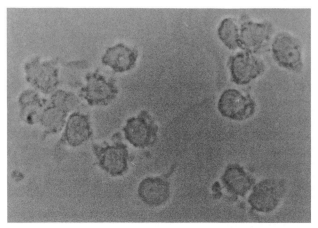

A

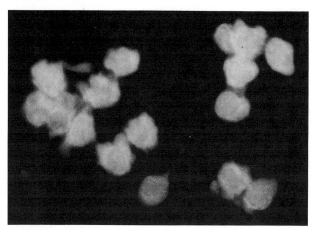

B

Burkitt's lymphoma cells—B cells with large nuclei and small cytoplasmic areas—viewed by a light microscope. A (*above*): Antibody directed against a cell-surface antigen, LMP, will bind to the surface of these cancer cells.
B (*above*): The same cells as at left. A fluorescent probe, here attached to the antibody, permits us to visualize it binding to the viral protein, LMP.
A (*below*): A similar test for antibody directed against the Epstein-Barr virus nuclear antigen (EBNA-1) demonstrates the viral-encoded nuclear protein in lymphoma cells in culture. It can also be used to detect the presence of antibody in people. B (*below*): The same cells as at left, demonstrating the presence of this nuclear antigen by means of a fluorescent probe attached to the appropriate antibody.

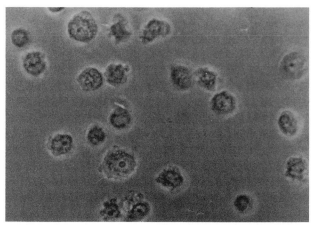

A

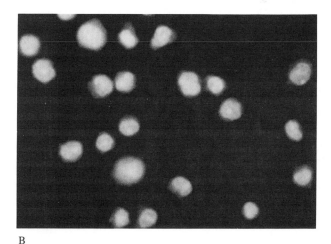

B

cells, stimulating polyclonal B-cell growth, but the killer T cells fail to function, and the uncontrolled replication of these B cells (a lymphoma) kills the host. Surprisingly, the pathology of these tumors (their appearance and location) is quite different from that of Burkitt's lymphoma, despite the presence of the same virus in both cases. Such tumors are rare, but their occurrence strengthens the evidence for the role of EBV in some human cancers.

In most patients, the symptoms of infectious mononucleosis fade with time, and all returns to normal. But the hallmark of the herpesviruses is latency, the ability to persist in a noninfectious form. All human beings infected by EBV, whether they have suffered clinical symptoms or not, carry the DNA chromosome of EBV in a small percentage of their B cells. We can demonstrate this in an interesting experiment.

If we take several milliliters of blood from a donor, it is possible to fractionate the blood cells into different types: red blood cells, B cells, T cells, macrophages or monocytes, and so on. Let us place the B cells in culture medium, where they have a chance to replicate. Even though we supply everything needed for growth (vitamins, growth factors, amino acids, and so forth), most B cells—indeed, most cells of our body—either do not replicate or replicate for a limited number of divisions. Limited life span is a property of normal cells, as we saw in Chapter 3. However, a few B cells occasionally demonstrate the ability to divide forever, and an immortal cell line can be obtained from them, as was the case with the cells from the Henles' technician. Furthermore, in every case where the donor has recovered from infectious mononucleosis, these permanent B-cell lines express some EBV antigens. While it is uncommon to detect infectious virus in these lines, the viral genome, in the form of a circular DNA molecule, persists in every B cell that replicates in the culture. This EBV DNA encodes a subset of mRNAs, and the proteins (EBNA, LMP) that they synthesize immortalize the B-cell lines in culture.

This is a little startling. In Chapter 3, we learned that a tumor virus can immortalize a nor-mal cell in culture. Now we find out that most of us carry a few B cells that contain EBV DNA in a latent state—which, under the correct circumstances, can immortalize these B cells for growth in culture. Clearly, cell growth in culture and in the living body shows differences, and not the least of these is the presence of killer T cells in the body that may well help to keep down EBV-directed growth of B cells.

Does this description of infectious mononucleosis and EBV explain how this virus contributes to Burkitt's lymphoma? How can a virus that causes infectious mononucleosis all over the world cause Burkitt's lymphoma in a limited area of Africa and New Guinea? EBV is indeed a human tumor virus, but it is clearly not carried by insects in Africa, so most scientists who have thought about this problem suspect that additional factors must be involved in causing Burkitt's lymphoma. The apparent EBV contribution rests on several observations:

1.   EBV DNA is found in the great majority of B-cell lymphoblastoid tumors of Burkitt's-lymphoma patients. That is a constant association between the two variables.

2.   EBV is the cause of infectious mononucleosis, which can result in B-cell lymphomas in immunosuppressed patients. Although B-cell lymphomas and Burkitt's lymphoma are distinct, these relationships are provocative.

3.   The inoculation of EBV into a marmoset (a small primate) induces a B-cell lymphoma, and EBV DNA is found in all these tumor cells. In all cases, a subset of EBV genes—the EBNA antigens and LMP—is expressed in these cells. It can be shown that some of these viral-encoded proteins are responsible for altering growth patterns and conferring immortality upon the cells.

4.   Finally, EBV has been shown to be closely associated with another human cancer, nasopharyngeal carcinoma (a cancer of the epithelial cells of the nasal and oral cavities). This disease also has a peculiar geographic localization, occurring in most cases only in southern China. In this group of Chinese people, EBV replicates in the oral cavity for many years. The cancer cells all harbor the EBV

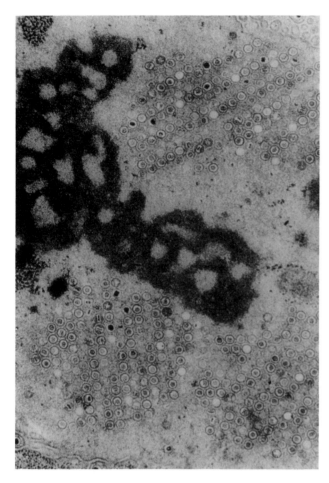

EM of herpesvirus particles, closely packed in an infected cell; these correspond to the characteristic virions harbored in the cells of patients with Burkitt's lymphoma.

chromosome in a circular DNA form, much as do B cells from infectious mononucleosis and B cells from Burkitt's lymphoma. While the evidence is convincing that EBV plays a role in both Burkitt's lymphoma and nasopharyngeal carcinoma in southern China, some additional factors must be involved in the creation of these cancers. Certain geographical or genetic variables may predispose particular groups to develop cancers when exposed to EBV, while other groups develop infectious mononucleosis.

Epidemiologists have searched for these additional factors. For example, in southern China, nasopharyngeal cancer is hundreds of times more common than in the United States. Furthermore, in southern China these cancer cells harbor the EBV chromosome, whereas it is very rare for the EBV chromosome to appear in nasopharyngeal cancer cells from patients in the United States. Among the possible variables in China and the United States are genetic differences between the people with these cancers and environmental differences such as dissimilar food, drink, or pollution. To discriminate between these factors, individuals from southern China who had emigrated to the United States were followed with respect to their incidence of nasopharyngeal cancer. While the incidence of this cancer declined in these immigrants, it did not approach the low rate in other Americans. That suggested that some environmental and some genetic factors may contribute to this EBV-induced cancer. It can be appreciated that such studies are not perfect; for example, most first-generation Chinese who live in the United States do not stop eating the same food they enjoyed in China—the environment is different, but not so different.

Burkitt's lymphoma in Africa appears to have a clearer environmental component. Whether that is the endemic malaria seen in the same regions as Burkitt's lymphoma or some other parasitic disease is not yet clear. The early age at which Africans are infected with EBV, the high rate of infectious diseases and parasites they suffer, the status of their immune systems under these conditions, and even nutritional variables could all contribute, along with EBV, to the appearance of Burkitt's lymphoma. It is odd that the very observation that started the search for a human tumor virus—that this cancer occurred in a region where virus-carrying mosquitoes could breed—is still unexplained.

More recent studies with Burkitt's lymphoma have demonstrated that the presence and expression of EBV DNA in infected cells may be necessary, though not sufficient, for this tumor to arise. Almost all Burkitt's lymphomas and several different B-cell tumors also contain a chromosome translocation: that is, portions of two of the forty-six chromosomes normally found in every human cell have fused together and formed a single abnormal chromosome. The fusion of chromosomes creates problems similar to those we would expect from splicing together two computer tapes from different spools: confusion occurs at the junction of the sites where the different sets of information are spliced. In each case of Burkitt's lymphoma, the same gene is found at the junction or break point in these chromosome translocations. As a result of the mixing of signals, this gene (which is called the myc gene) is expressed at abnormally high levels. This excessive expression of the myc gene contributes to the cancer in ways we do not quite understand yet.

The same myc gene has been shown to play a role in a number of tumors, and we will meet it again when we examine the way in which retroviruses cause cancer. The myc gene is one of the fifty or so potential oncogenes (cancer-causing genes) that we have in our chromosomes. The normal function of these genes is important for the regulation of our growth and cell replication. When altered by mutation (as when a translocation makes more myc protein), they can contribute to human cancers. Thus, the genesis of Burkitt's lymphoma is complicated; it seems to require EBV, a myc gene translocation, and an unknown environmental factor, perhaps transmitted by insects. This list is probably not complete.

Cytomegalovirus and Herpesviruses Types 6 and 7

The fifth, sixth, and seventh examples of human herpesviruses—extremely common, yet little studied and understood—are cytomegalovirus (CMV) and the newly isolated human herpesviruses types 6 and 7. CMV, like the other herpesviruses, is spread by person-to-person contact; indeed, almost everyone is infected with this virus, which most often is not very pathogenic (though it is involved in a small number of cases of hepatitis and infectious-mononucleosis-like syndromes). Recently, however, it has become clear that intrauterine CMV infection is a serious problem. About 30,000 to 35,000 infants (1 percent of all live births) annually in the United States have congenital CMV infections. In a variable percentage of these children, defects in the central nervous system and in perception are detected over the first two years of life. CMV infections are probably a major cause of birth defects and learning disabilities—certainly the major viral contributor to birth defects now that there is a vaccine against rubella (German measles) virus.

After primary CMV infection, the virus becomes latent in the white blood cells of the body, although exactly which cell types it resides in remains unknown. The presence of latent CMV can pose a danger in some of the newest medical procedures, such as transplants where bone marrow (which synthesizes all the blood cells) is taken from one person and placed into another. CMV in these transplanted bone-marrow cells often activates, infecting the new host; CMV-induced pneumonia and hepatitis are serious and frequent complications of these procedures.

A way to prevent CMV infection in utero and after transplants is badly needed. Acyclovir does not prevent CMV replication effectively, but a related drug, gancyclovir, has been used with success to treat CMV retinitis, an eye infection that is becoming common in patients with AIDS. Herpesvirus reactivation occurs often in AIDS patients. Herpesvirus type 6, first identified in an AIDS patient in 1986, is a ubiquitous virus that appears to cause a common childhood rash. It remains latent in adults and can be isolated from white blood cells. Herpesvirus type 7 has been identified only very recently.

Although this chapter has focused on the human herpesviruses, most animals also have herpesvirus infections that often parallel the events described here. Among the most interesting is the Marek's disease herpesvirus, which causes a T-cell lymphoma in chickens; the cancerous T cells all contain the viral genome, as is the case with EBV. Of considerable interest is the related herpesvirus of turkeys, which is not pathogenic in chickens. This turkey virus has been used as a vaccine for chickens, because it prevents the Marek's disease virus from inducing chicken T-cell lymphomas. This is the first example of a successful vaccine against cancer. Given this evidence for the causative role of herpesvirus in this cancer, it is to be hoped that this vaccine model can be repeated for some of the human cancers with viral etiologies.

The human herpesviruses have developed a complex set of interactions with their host. They have found a way to infect us and remain with us for a lifetime, evading our immune defenses and evolving complex mechanisms for reactivation. Both the neurotropic herpesviruses (HSV-1, HSV-2, varicella-zoster virus) and the lymphotropic herpesviruses (EBV, CMV, HSV-6) have learned to reactivate in response to signals outside their immediate environment—whether trauma, immunosuppression, sunlight, sexual stimulation, or hormonal changes. Latency, the hallmark of the herpesviruses, remains one of the mysteries of virology. To understand it will be a challenge for virologists of the next generations.

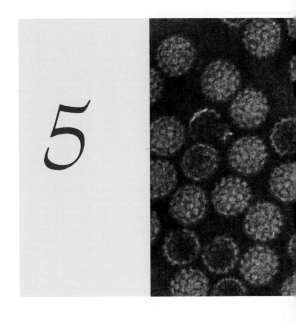

The DNA Tumor Viruses

When you are a student . . . science is new and wonderful and everything that is known is accepted as though it had always been known.

It is much later, after years of research, of failed experiments and successful experiments, of false leads and surprising results and true moments of insight, that one comes to realize how hard won is each nugget of knowledge; how each advance relies on all the cumulative advances of the past; how each experiment is part of the ongoing stream of science, part of the growing treasure of human knowledge that has brought us out of the darkness.

Robert L. Sinsheimer
June 21, 1990

The Epstein-Barr virus is a good example of a DNA tumor virus. It is intimately associated with the cancer cells from Burkitt's lymphoma and nasopharyngeal carcinoma. These tumor cells each contain a copy of the viral chromosome, in the form of a circular DNA molecule that resides in the cell nucleus. The viral chromosome expresses some of its genes, producing proteins that appear to contribute to the abnormal cell replication patterns we recognize as cancer. A number of other DNA viruses have similar capabilities, and although the details of how they express these may differ, the results are the same—they induce what, in this context, we can define as cancer: the abnormal growth or replication of cells.

All cells are programmed to respond to signals to begin dividing and signals to stop dividing; as we have seen, normal cells appar-

Left: Cottontail rabbits, from J. J. Audubon's *Viviparous Quadrupeds of North America.* A papilloma virus that produces warts in these rabbits—its normal hosts—induced carcinomas when inoculated into another rabbit species. It was eventually identified as the first known DNA tumor virus.

Above: EM of human papilloma viruses, showing the capsomere subunit structure and the icosahedral symmetry of the nucleocapsids. The broken particles do not contain viral DNA.

ently can undergo only a limited number of divisions in culture. Cancer cells arise either because the signal to replicate is constantly on or because they do not obey the instruction to stop dividing; in some cases, the signal to stop growth is lost via a mutation in the cell's chromosome. When the program for regulating cell division goes awry, cells replicate at inappropriate times and in inappropriate places, and they no longer have a limited life span. Viruses can encode the genetic information to produce proteins that change normal cells into cancer cells. For the DNA tumor viruses, this was first shown by Richard Shope in 1932 and 1933.

The Discovery of DNA Tumor Viruses

In the early years of the Depression, several hunters in the hills of Cherokee, Iowa, noticed that the local rabbit population had hornlike protuberances on their bodies. The hunters brought some of these "horned" or "warty" rabbits to Richard Shope, who quickly demonstrated that a filterable agent was present in the soluble extracts of the warts and that this agent could induce warts in other rabbits.

Warts are benign tumors. They are classified as tumors because they arise from the abnormal growth of cells in the basal (bottom) portion of the skin. As normal, maturing skin cells continue to divide, they synthesize a variety of tough, fibrous proteins called keratins that, as the predominant proteins in the cells, make skin tough and protective. As the keratins accumulate, they eventually kill the cell, and we continually slough off the dead skin cells. If skin cells in a particular area divide excessively, they overproduce keratins and a wart, or horny region of the skin, develops. What Shope found was a rabbit wart virus that gets into the basal layer of skin cells and promotes active cell division. The dead cells in such a wart are full of viruses, which reinfect other rabbits. This virus has been named Shope papilloma virus (papilloma is a clinical term for wart).

The next act in this story was carried out by Peyton Rous, who received a sample of the papilloma virus from Shope. In his laboratory at the Rockefeller Institute in New York City, Rous kept only domestic rabbits purchased from a breeder. All of Shope's rabbits were wild—a different species from the domestic rabbit. Nonetheless, Rous inoculated the Shope virus into his domestic rabbits, where the virus replicated poorly and seemed to disappear. Its infectivity was completely lost. (This is an example of a phenomenon we have seen before: a virus will grow poorly in a new host species, unless or until it becomes adapted to that host.)

But, some time later, these rabbits were observed to develop carcinomas of the skin. A carcinoma, unlike a wart, is a malignant tumor. The skin cells produce abnormal keratins and continue to divide in an uncontrolled fashion. In a wart, the tumor cells produce the normal skin-cell keratins and eventually die. The cells in a carcinoma do not differentiate into terminal skin cells—they do not die. This is a far more serious disease.

Rous tried to fulfill Koch's postulates by isolating infectious virus from these cancer cells. In every case, he failed. The virus had disappeared.

Subsequent studies showed that every cell of this carcinoma contained a portion of the papilloma viral DNA integrated into the chromosome of the host cell. Unlike EBV, which leaves a free, circular viral chromosome in the nucleus of the cancer cell, other DNA tumor viruses initiate a cancer by inserting their DNA into that of the host cell, so free viral DNA is not found and infectious viruses cannot be produced. The malignant tumor originates in one event: a single integration of the viral DNA into the chromosome within a single host cell. All the cancer cells are descendants of this one cell, and each of these malignant descendants contains the viral DNA, inserted into the same place in the chromosome. Because this insertion is in the chromosome of skin cells, not the sperm or egg cells (the germ line), this cancer is not inherited by the offspring of the host. The viral DNA is present only in these cancerous skin cells: the tumor. In each cancer cell, furthermore, a portion of the viral DNA is transcribed into mRNA, which, in turn, makes a

Francis Peyton Rous.

subset of the viral proteins. These proteins have been shown to cause the abnormal cell growth. Because the expression of these viral genes results in a cancer, they have been called oncogenes, or cancer-causing genes.

When a viral-encoded protein is produced in a cancer cell, it sometimes becomes exposed to the host's immune system, either because the cell has died and disintegrated or because a portion of the protein has moved to the cell surface. The immune system recognizes these viral proteins as foreign and commonly produces antibodies directed against them; killer T cells can even attack and destroy cancer cells, under the appropriate conditions. Virologists often use the antibodies taken from animals with tumors to detect foreign viral proteins in the tumor cells. Because the viral-encoded proteins are targeted by antibodies, they have come to be called tumor antigens. Recall that the Henles used the antibodies from patients with Burkitt's lymphoma to detect EBV-encoded tumor antigens in lymphoid cells derived from that lymphoma. Tumor antigens in rabbits demonstrate parallel functions.

At this point, we might ask: If the immune system responds to foreign viral proteins by producing antibody and killer T cells, why do these tumors occur at all in people or animals? There is no simple, clear answer to this question. The tumor appears to get a head start, producing dividing cells that reach a critical mass before an immune response can be mounted. The immune system may be providing too little, too late. To test this idea, consider a mental experiment much like the one we used to demonstrate immunity against virus infections. If we have two mice and immunize one with killed tumor cells, we will find that this mouse makes antibodies and directs killer T cells against the viral-encoded tumor antigens on the surface of the cancer cells. If we now challenge these two mice with live cancer cells, they will be rejected by the immunized mouse (no tumor) and accepted by the nonimmunized mouse (a tumor will form). Prior exposure can permit the immune system to reject a tumor. We now inject the serum antibodies from mouse 1 into a nonimmunized mouse and its killer T cells into a second nonimmunized mouse. If we challenge these two mice with an injection of live tumor cells, only the killer T cells will offer passive immunity against a cancer. The antibodies, while useful in identifying the viral proteins as tumor antigens, cannot protect against the formation of a cancer. The arm of the immune system that defends us against cancer is cellular immunity—the killer T cells—not humoral immunity (that is, antibodies).

Over the years since the mid-1930s, when Shope and Rous made these original observations, several additional groups of viruses have been shown to be DNA tumor viruses, In each case, the paradigm remained as Rous and Shope had demonstrated: a virus commonly isolated from one host animal could induce a tumor in a different species.

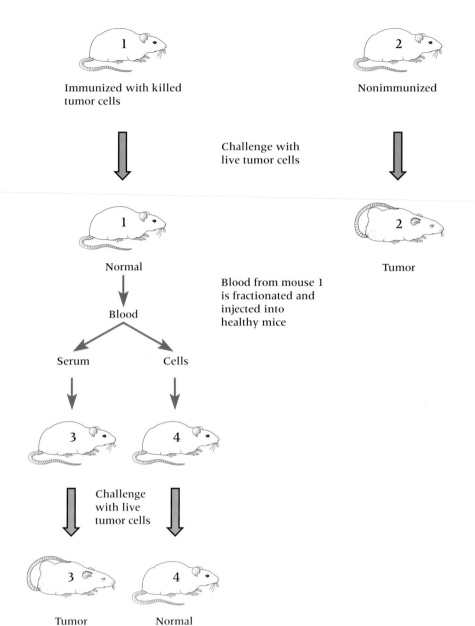

Immunized with killed tumor cells

Nonimmunized

Challenge with live tumor cells

Normal

Tumor

Blood

Blood from mouse 1 is fractionated and injected into healthy mice

Serum

Cells

Challenge with live tumor cells

Tumor

Normal

A demonstration of the role of cellular immunity in cancer. This experiment shows that killer T cells can passively protect against a tumor; antibodies cannot.

The tumor cells did not contain detectable infectious virus but had instead, in every cancer cell, a portion of the viral chromosome integrated into the host DNA. Because the millions of cells in a tumor all displayed viral DNA integrated at the same site in a host chromosome, it could be determined that the tumor arose as a clone of cells from one single ancestor—it was a rare event. Tumors arising in different animals inoculated with the same virus each had a distinct integration site in a chromosome, although within each individual tumor, the integration site was identical in all cells. All of the cancer cells produced a viral mRNA that, in turn, synthesized viral oncogene products. These viral proteins elicited an immune response that generated antibodies to identify tumor antigens and mobilize killer T cells that could, under some circumstances, protect against a viral-initiated tumor.

Viruses that fulfill this paradigm have been discovered in monkeys (simian virus 40, or SV40) and humans (the BK and JC viruses and the adenoviruses). DNA tumor viruses that initiate tumors in their natural hosts have also been described: in mice, polyoma virus, and in humans, the human wart viruses and human papilloma viruses types 16 and 18, which are associated with anogenital carcinomas.

Simian Virus 40 (SV40)

SV40 was first detected and isolated in a most unusual set of circumstances. In the middle of the 1950s, when the inactivated Salk polio vaccine had proved to be efficacious, the pharmaceutical companies that had licenses to produce it had to find ways to grow large batches of poliovirus from which to prepare a vaccine. Poliovirus replicates only in humans and monkeys. The federal agencies overseeing vaccine production did not like the idea of using a permanent human cell line, grown in culture, to replicate this virus; such an immortalized line could harbor contaminating viruses, even cancer-causing viruses. That left the use of monkey cells—in particular, *primary* monkey cells (that is,

cells derived from animals directly, rather than immortalized for growth in culture). In the end, primary cells isolated from the kidneys of *Rhesus* and *M. cynomolgus* monkeys were chosen to grow poliovirus for the millions of doses of Salk vaccine.

After the poliovirus replicated, killing these cells, the filtered solution was treated with formalin to inactivate the virus. These inactivated doses of virus were then tested to determine: (1) Did they elicit an immune response? (2) Did they cause any disease when injected into monkeys (was there any residual live virus)? (3) Did the virus replicate when incubated with monkey cells in culture? (4) Was any disease produced when the vaccine was inoculated into various other animal species— mouse, hamster, rabbit, and so forth? (The testing of each lot of a vaccine for toxic effects is good pharmaceutical practice.)

All the short-term tests, based on examination of the cell cultures over a period of weeks, were negative. The immune response was good. By this time, the test vaccine had been given to thousands of people, and production was being scaled up for the introduction of a large vaccination program. It was at this moment that some of the hamsters in these experiments, which had been inoculated at birth with the inactivated poliovirus vaccine, began to show tumorous growths.

The latent period for tumor appearance—the time before it was first detected—varied between 130 and 327 days after inoculation; the average was 230 days. Out of 151 animals inoculated, 42 developed tumors. There was no infectious virus in these tumors, but serum from animals with tumors contained antibodies that detected tumor antigens; these antibodies were now used to detect a virus in the primary monkey-kidney cell cultures first employed to replicate the poliovirus. This monkey virus did not kill the cultured cells, but it did produce a strange pattern of vacuolation (vesicle formation) in them; it was named simian vacuolating virus 40 (SV40).

Once it was appreciated that some doses of the early poliovirus vaccine were contaminated with a monkey virus that could cause tumors in hamsters,

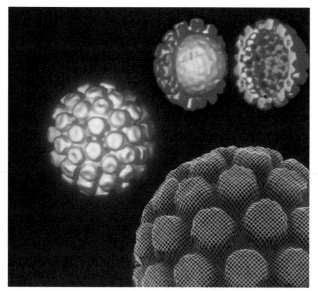

A   B

Computer-based three-dimensional reconstructions of SV40, from EM measurements. A: (Top) Cutaway views of the virus particle showing the inside of the capsid (red) and the capsid cut in cross section with the nucleosomes-viral DNA minichromosome. Surface representation of the virion (blue), viewing the twofold, threefold, and fivefold rotational symmetry axes. B: SV40 virus particles.

the people in whom these vaccine doses had been tested were followed in great detail. The formalin treatment that killed the poliovirus did not kill SV40, which was able to replicate in people who had been given the contaminated vaccine—SV40 was secreted in their feces and urine, and they made antibodies directed against it. But a small group of infants originally inoculated with this virus have been followed for over thirty years now, and to date no abnormalities or cancers have appeared in this group. When scientific investigators who had worked with large amounts of SV40 in their research laboratories subsequently died of cancer, tests established that no viral DNA was present in their tissues. No evidence to date associates this virus with cancer in the human population. Why SV40, a monkey virus, is able to produce

tumors in rodents but does not, apparently, do so in humans remains a puzzle.

The SV40 virus contains circular, double-stranded DNA that is 5243 nucleotide base pairs long. This DNA is packaged into a condensed virion core by being wrapped around the cellular histone octomers (nucleosomes). In this case, the four different cell-encoded histone proteins are used to package viral DNA in a manner very similar to the way host-cell DNA is packaged in the nucleus. Surrounding this core is an icosahedral shell made of three viral structural proteins (VP-1, VP-2, and VP-3), which are the late viral proteins. SV40 also encodes two early proteins, synthesized shortly after cell infection: these are called the tumor antigens, or T antigens. T antigen is a key orchestrator of the events leading to viral DNA replication.

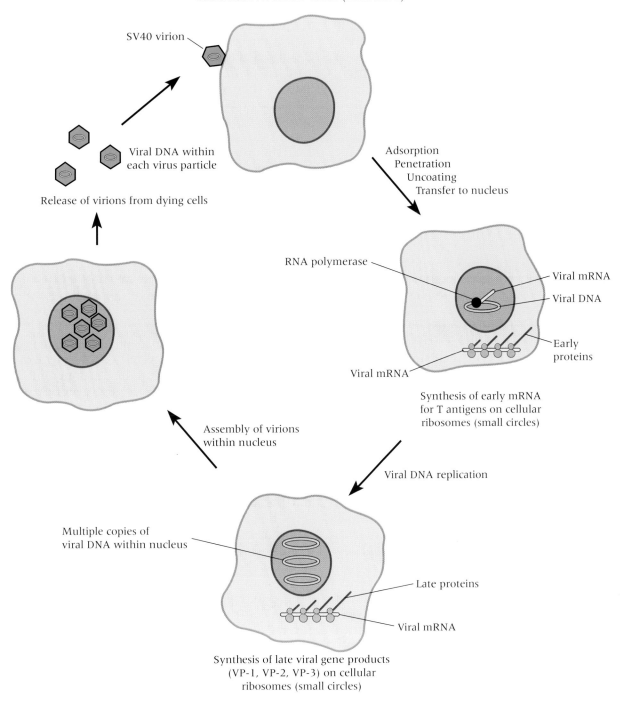

SV40 virion

Viral DNA within
each virus particle

Release of virions from dying cells

Adsorption
Penetration
Uncoating
Transfer to nucleus

RNA polymerase

Viral mRNA

Viral DNA

Early
proteins

Viral mRNA

Synthesis of early mRNA
for T antigens on cellular
ribosomes (small circles)

Assembly of virions
within nucleus

Viral DNA replication

Multiple copies of
viral DNA within nucleus

Late proteins

Viral mRNA

Synthesis of late viral gene products
(VP-1, VP-2, VP-3) on cellular
ribosomes (small circles)

Schematic diagram of the steps in the replicative cycle of SV40 in its normal host, the monkey—lytic infection.

The genes that code for the T antigens take up half the viral DNA, and their mRNAs are transcribed in a clockwise direction on the circular chromosome. The mRNAs for the late gene products are transcribed in the counterclockwise direction. This means that the coding sequences (the sense, or plus, strands) for the two types of protein are on opposite DNA chains. The start site for the early mRNAs (clockwise) and the start site for the late mRNAs (counterclockwise) are separated by a region of about 300 base pairs (bp). This region of the viral chromosome contains a number of regulatory nucleotide sequences that act as signals to promote or inhibit transcription. Also in this region is the starting point for SV40 DNA replication.

The virus attaches to monkey cells in culture by means of VP-1, but little is known about the cellular receptor recognized by this virion protein. The virus particles are apparently brought into the cell by endocytosis and transported to the nucleus, where they are uncoated, releasing the viral DNA. It is at this time that the regulatory region of the viral chromosome comes into play. Unlike the herpesviruses, which encode some of their own enzymes, SV40 must mobilize the equipment of the host-cell nucleus in order to replicate the viral DNA. The functions of the 300-bp regulatory sequences are crucial here, and we will examine them in some detail.

There are three distinct nucleotide sequences in this region of the viral chromosome: (1) About 30 bp before the start site for early mRNA synthesis is a nucleotide sequence, TATA, that directs the initiation of mRNA synthesis at the start site. (2) Between 40 and 100 bp before the T-antigen mRNA start site is a 21-bp sequence, repeated three times. (3) About 110 bp before the early-gene mRNA start site is a 72-bp nucleotide sequence, repeated twice. The TATA sequence and 21-bp repeats are termed the promoter: these elements determine the direction and start site of transcription. The 72-bp repeat is termed the enhancer: it increases the level of mRNA transcription from the start site chosen by the promoter. These three signals, making up the enhancer and promoter elements, function by attracting or binding cellular proteins called transcription factors (transactivators); these, in turn, attract the cellular RNA polymerase enzyme to the T-antigen gene, where it begins transcription of the mRNAs for T antigen. (The enhancer-promoter signals on the SV40 chromosome are very similar in structure and nucleotide sequence to the enhancer-promoter elements on the host-cell chromosome, whose job it is to regulate the genes of the cell.)

The availability of these transcription factors determines if a gene will make mRNA. Some transcription factors are tissue-specific, so some genes can be expressed (that is, make mRNA) only in a particular tissue. SV40, moreover, cannot produce virus if the transcription factors in an infected cell fail to recognize the SV40 chromosome. Sometimes mutations occur in the nucleotide sequence of the viral enhancer that change the host range of a virus—the cells in which it can replicate. Such host-range mutant viruses can cause new disease patterns.

Once the cellular transcription factors bind to the SV40 chromosome, the cell's RNA polymerase initiates clockwise transcription, making mRNA across the early-protein (T-antigen) genes. A specific nucleotide sequence signals for the mRNA to be cut and terminated at a precise spot on the chromosome. This mRNA, now 2470 bases long, is further processed—cut and spliced—into two alternative forms. This processing of the mRNA transcript removes internal nucleotide sequences, creating mRNAs for the two T antigens of SV40.

The completed mRNAs pass from the nucleus of the infected cell into the cytoplasm, where the two T antigens are synthesized on the host-cell ribosomes. In the amino-acid sequence of the large T antigen resides a signal or ZIP code that directs this protein to return to the cell nucleus. There it recognizes a particular nucleotide sequence that appears several times in each of two places (called site I and site II) in the 300-bp regulatory region of the viral chromosome. Both sites bind T antigen specifically (like the lambda repressor we considered in Chapter 2), but they have different functions.

When T antigen binds to site I, it prevents RNA polymerase from attaching to the gene for T antigen. T antigen, then, acts as a repressor for its own

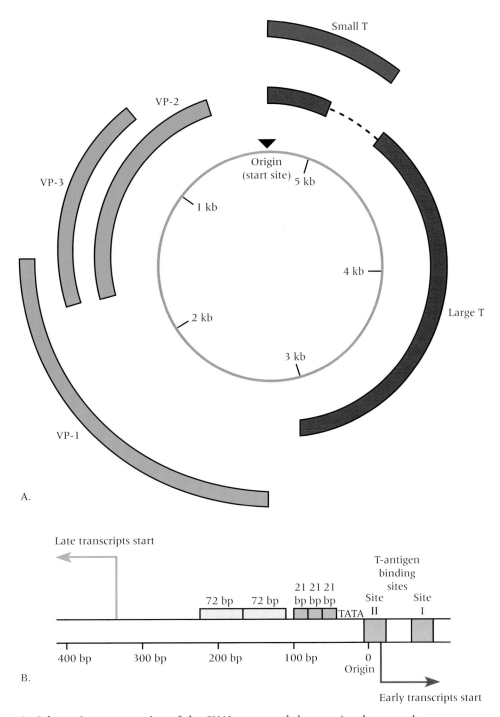

A: Schematic representation of the SV40 genes and the proteins they encode. (kb = kilobases, 1000 bases) B: An enlargement of the region between the start of early and late transcription.

gene, so that the level of T antigen in a cell is kept constant. The binding of this antigen is concentration dependent—when too little T antigen is present, none will bind to site I. Consequently, more T-antigen mRNA is made, resulting in more T-antigen protein. When the amount is increased to a level sufficient to bind to site I, synthesis of the gene product stops, keeping the protein level within the specific range that is critical for T antigen to function optimally. T antigen is said to autoregulate.

The second function of T antigen is to bind at site II, initiating the replication of viral DNA. At this site, T antigen actually separates the two strands of viral DNA and starts the synthesis of new strands at a specific region: the origin of DNA replication. Synthesis of new DNA strands proceeds from this origin both clockwise and counterclockwise (bidirectionally) to a terminus 180 degrees around the circle. As the viral DNA is synthesized, it is packaged into cellular nucleosomes that condense and compact the viral DNA, still within the cell nucleus.

After DNA replication, the late genes are transcribed and expressed; this also requires signals from the regulatory region of the viral DNA. The three late mRNAs, made from thousands of newly replicated DNA templates, synthesize their proteins in concentrations hundreds of times higher than that of the T antigens, since the structural proteins are needed in large amounts to produce the thousands of virions in an infected cell. Assembly of the capsomeres from VP-1, VP-2, and VP-3 proceeds spontaneously in the nucleus. Viruses escape when the dying cell falls apart, exhausted.

To replicate its DNA, SV40 encodes only one protein: the large T antigen. The limited amount of genetic information in SV40 (5243 bp) constrains this virus to rely upon the host cell to supply enzymes like thymidine kinase, which make nucleotide precursors, and enzymes like DNA polymerase, which synthesize DNA. The virus has evolved a mechanism to stimulate increased levels of activity in cellular enzymes by inducing the cell to enter its own replication cycle, synthesizing cellular DNA and, in some cases, even going on to cell division. By contrast, HSV-1 (with 152,000 base pairs) en-

codes for its own thymidine kinase and DNA polymerase, does not stimulate host-cell DNA replication or cell division, and—unlike SV40—is not a tumor virus. Several experiments have shown that the SV40 large T antigen is responsible for (1) the stimulation of the cellular thymidine-kinase activity in an infected cell (mutant T-antigen cells do not do this) and (2) the stimulation of cellular DNA synthesis in the infected cell. SV40 T antigen not only helps SV40 DNA to replicate, but compels the cellular DNA to replicate.

SV40 and the Transformation of Cells in Culture

The life cycle detailed in the preceding section describes the fate of an SV40 virus and the cell it infects in its normal host, the monkey. The virus reproduces, the monkey cell dies, and infection spreads. What happens to the virus and its host cell is quite different in hamsters and other animals that develop cancers when infected with SV40. In this case, the SV40 T antigen is synthesized, but the viral DNA does not replicate and no late proteins are made. The cell is stimulated to produce its own DNA replication enzymes, in large amounts. Some of these enzymes synthesize nucleotide precursors of DNA, and others, such as DNA polymerase, manufacture DNA. The signals to enter into DNA synthesis are given, and the cell duplicates its genome. In a rare cell, the SV40 chromosome integrates into the cellular DNA and transcribes T-antigen mRNA. When T antigen is produced all the time, the cell is continuously signaling for growth and replication. A cancerous cell results.

The ability of T antigen to trigger replication of the cellular genome suggests how SV40 may create the uncontrolled cell division that leads to malignant hamster tumors, but it is difficult to prove that the SV40 T antigen acts to promote such tumors by stimulating cell division directly. When newborn

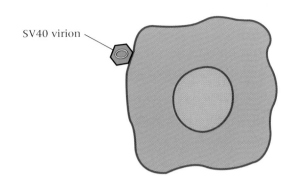

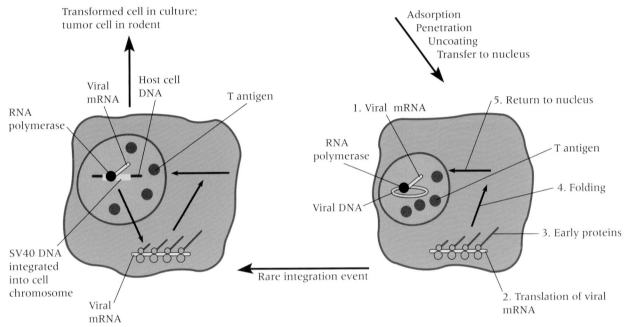

Transformed cell in culture;
tumor cell in rodent

Adsorption
Penetration
Uncoating
Transfer to nucleus

Viral mRNA

Host cell DNA

T antigen

1. Viral mRNA

5. Return to nucleus

RNA polymerase

RNA polymerase

T antigen

4. Folding

Viral DNA

3. Early proteins

SV40 DNA integrated into cell chromosome

Rare integration event

2. Translation of viral mRNA

Viral mRNA

Cell with integrated SV40 DNA; continuous synthesis of T antigen
1. Stimulates levels of host-cell enzymes involved in DNA replication (thymidine kinase and DNA polymerase)
2. Stimulates synthesis of cellular DNA and nucleosomes

Synthesis of T antigen on cellular ribosomes (small circles), folding of protein, and transport back to the nucleus

Schematic diagram of the impact of SV40 infection upon cellular functions in hamsters or other rodent hosts—the transformation cycle.

hamsters are inoculated with SV40, the virus disappears into a few cells of the body. Months later, a tumor appears near the injection site. What happens in the time between these two events cannot be studied in detail. There is a clear need to leave animal studies and examine the activity of the virus directly, by observing the events that occur as SV40 interacts with a cell in culture and changes the cell's growth properties. Could we develop an assay in cell culture that reflects the events eventually leading to an animal cancer? Even if we could, how would we recognize that it mimicked the viral-cellular processes that occur in an animal?

One way to answer these questions is to examine cells in culture that have recently been obtained from a normal host and then contrast their properties with those of cells derived from an animal cancer. (We did this in Chapter 3 for one cellular property, showing that normal cells have a limited life span in culture, whereas cancer cells are immortal in culture. We also saw that SV40 infection of normal human cells in culture can result in the appearance of rare clones of immortalized cells.)

There is a remarkably good coincidence of the events occurring in tumors and in cell culture. In both cases, we observe the integration of viral DNA into the cellular chromosome, followed by the expression of early mRNA and the synthesis of T antigen. Furthermore, a mutant SV40 virus exists that produces a viral-encoded T antigen that functions normally at 32°C but fails to function at 39°C. This is called a T-antigen temperature-sensitive mutation. Because the SV40 T antigen is required for viral DNA replication, a cell containing this mutant will produce virus at 32°C (the permissive temperature) but not at 39°C. When normal rat cells in culture at 32°C are infected with this T-antigen temperature-sensitive SV40 mutant, rare cells produce permanent cell lines that can grow at 32°C for long periods of time. If we now shift the temperature to 39°C, the T antigen fails to function, and the cells stop growing. They have a limited life span without the functional T antigen. Clearly, T antigen is necessary, although it may not be sufficient, for the immortalization of cells in culture. Over the years,

a number of other assays have been developed to recognize the differences between normal cells in culture and cancer cells in culture. It will be useful to review these assays: collectively, they define the growth-altering properties of the cancer-causing viruses.

The great majority of cells require signals from the extracellular environment to stimulate their replication and division. In most cases, these signals come in the form of proteins called growth factors, which specifically bind to growth-factor receptors on the plasma membrane and trigger them to transmit a signal across the membrane and into the cell. These messages then follow what is called a signal transduction pathway to the nucleus, where they promote the transcription of new genes, whose functions carry out cell growth and division. (Some of those genes make thymidine kinase and DNA polymerase, for example.) The SV40 large T antigen has learned to trigger this signal transduction pathway, turning on the cell and promoting division.

As cells grow in culture, they often require growth factors, which are supplied by adding serum proteins derived from young animals. In practice, normal cells can grow in medium supplemented with 10 percent (one part in ten) fetal calf serum. If we reduce the level of such growth factors to 1 percent (one part in one hundred), the normal cells fail to replicate in culture; they simply stop dividing. Interestingly, cells derived from cancers that replicate in culture can often grow normally in 1 percent fetal calf serum. Cancer cells, then, are either independent of or less dependent upon growth hormones to stimulate their division.

Normal cells, growing and replicating in 10 percent serum on the surface of a culture dish, divide and form a colony. One cell often lies next to its neighbor, touching it, but cells never grow on top of each other. Cells in the middle of the colony, surrounded by neighbors on all sides, usually stop dividing, while cells at the edge, with space on an outside surface, continue to divide. This has been termed contact inhibition of cell growth. It results in a colony of cells that grow only in two dimensions on a flat surface, never upward in multiple, three-dimensional layers. The colony, accordingly, looks flat—like a thin pancake.

Cancer-cell growth is not contact inhibited. At the center of a colony, cancer cells divide in the third dimension, continuing to replicate one on top of the other. This produces a colony many cell layers thick. The cancer cells can produce foci of cells, where each focus is a multicellular colony composed of cells replicating up as well as across.

Normal cells in culture can divide only when they have a surface to grow on. In practice, the culture dish provides a plastic or glass surface for the flat, pancakelike colony, one cell thick, to adhere to. If normal cells are placed in a suspension of agar where there is no surface for them to attach to, they fail to divide. Cancer cells, on the other hand, divide very well whether on a surface or in a suspension; that is, cancer cells are anchorage-independent for their growth.

When normal cells from an animal are injected back into that animal, they replicate normally and

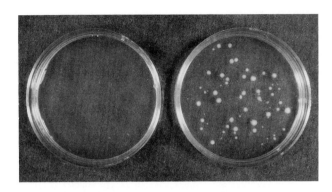

Single cells were placed in agar suspension in culture dishes and allowed to grow for two weeks. The normal cell (left) failed to grow; no colonies derived from it can be observed. In contrast, the transformed cell (right) produced colonies (white spots), demonstrating its ability to replicate under these conditions.

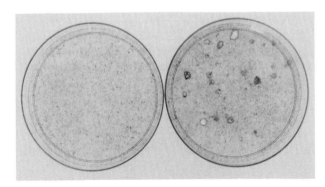

Uninfected cells (left) and cells infected with SV40 (right) growing on the surface of culture dishes. Each set has been treated with a stain that makes multiple layers of cells in a transformed colony appear dark. The uniform lightness of the culture dish on the left indicates that the culture is normal: only one cell layer thick. The infected culture dish is a mixture of normal cells and darkly stained colonies of transformed cells. Each transformed cell colony, a clone derived from a single cell containing the SV40 genome integrated into the cellular DNA, forms foci and expresses the SV40 large T antigen.

do not form a tumor. In many cases, when cancer cells are injected back into an animal, they do form a tumor. In practice, one can employ either an immunosuppressed animal or a normal, immunocompetent animal for this test of tumorigenic cell potential. Because tumor-specific antigens are recognized by killer T cells, which can attack tumors, tumors tend to be more readily produced in immunosuppressed hosts. Normal cells are not tumorigenic in either type of host.

Finally, it is often observed that cancer cells and normal cells have very different morphologies—that is, size, shape, organelles, and other visible features. Virus-induced cancer cells are usually smaller, more spherical, and more refractile to light passing through them than are normal cells, and they are attached poorly to the surface of the culture in comparison to their normal counterparts. These morphological differences can be very striking.

These six criteria—immortalization of cell growth; growth in low serum; loss of contact inhibition (foci formation); loss of anchorage dependence; tumorigenic potential; and altered morphology—permit us to distinguish between normal cells and cancer cells in culture. In reviewing these criteria, it is interesting to observe that in all cases

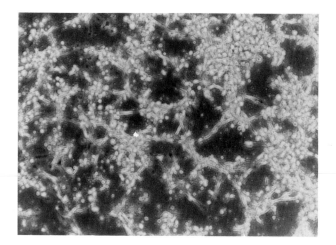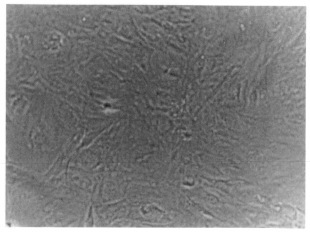

Light-microscope photographs of the morphologies of normal cells in culture (right) and transformed cells in culture (left). Normal cells are larger and less refractile to light; they often orient themselves in parallel array adjacent to one another.

the cancer cell can do something the normal cell cannot do. Because we cannot be sure that one or all of these criteria really reflect similar mechanisms involved in the initiation of cancer in a living animal, we have adopted a special term for cells that are not normal in cultures. Cells showing one or more of these criteria for abnormal growth in culture are said to be *transformed*.

We can now ask additional questions about the transformation of cells in culture. Can a cell be transformed in terms of one or two of these criteria and not the others? For example, can a cell be immortalized and grow in low serum, but fail to form foci, grow in agar, or be tumorgenic in animals? Is the transformed state an all-or-none property, or can it be partitioned?

To answer these questions, we can take normal cells cultured directly from a mouse and infect them with SV40 (recall that the virus's normal host is a monkey). SV40 can enter mouse cells and transform them, but for some reason it does not replicate in these cells. Mouse cells can be termed nonpermissive for SV40 replication, although the virus does express its tumor antigens in these cells.

To carry out our experiment, we can employ each of the six criteria for transformed cells to select for rare transformation events in a background of lots of normal cells. In low serum, for example, only the rare transformed cell will grow and duplicate itself, forming an immortalized cell line. If several independently obtained immortalized cell lines, derived from distinct transformation events, are tested for growth in low serum, some will replicate and some will not. When those lines that could replicate in low serum are further tested for loss of contact inhibition, some will prove to be transformed for this property, and some will not. Similarly, if these cell lines are tested for their ability to grow in agar or form tumors in animals, only some will have the ability to fulfill all these transformation criteria.

Independently transformed cell lines all have the SV40 DNA integrated into their genomes and express the T antigen, but some cell lines are transformed for all six properties, some for only three, and others for only one. However many of these transformation criteria are fulfilled, the SV40 large T antigen is necessary to maintain those particular cell properties. Yet the presence of the SV40 large

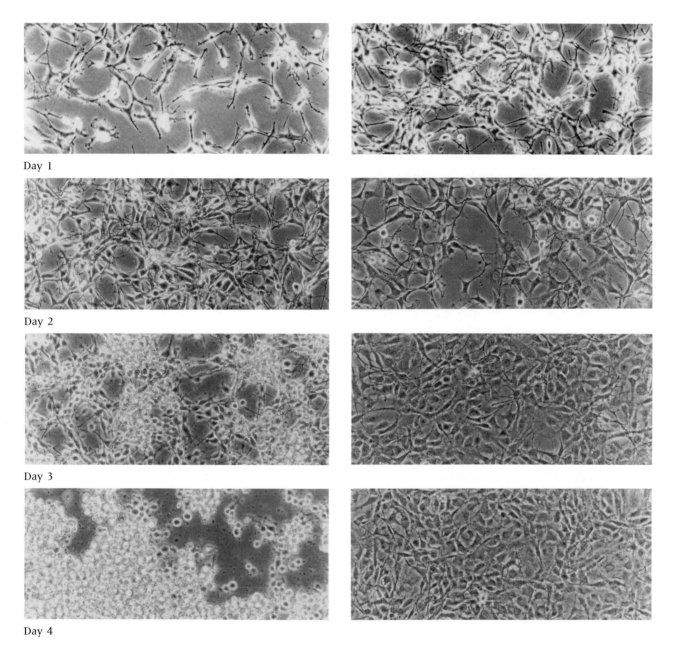

Day 1

Day 2

Day 3

Day 4

Here, rat cells regulated by the expression of an oncogene demonstrate two of the six criteria for cell transformation. When the oncogene is turned off in these cells (right), both the transformed cell morphology and the abnormal ability to grow in colonies on top of each other revert to normal after two or three days. If the oncogene continues to be expressed (left), the cells remain transformed for these properties

T antigen is not sufficient to ensure the fulfillment of these criteria. Something in the host cell must also affect the growth properties of the transformed cells. The heterogeneity of transformation demonstrated by different cells in culture probably, in fact, reflects the observed heterogeneity of cancer-cell properties in living animals and the complexities of actual malignant tumors.

We may take from these observations several key conclusions, which we will continue to develop with additional groups of DNA tumor viruses. First, a set of DNA viruses can induce tumors in animals, either in their natural hosts or in other host species. Second, the tumor cells will contain the viral genome, or part of it, integrated into the host-cell chromosome; this viral DNA will express tumor antigens, which play a key role in causing the cancers. Third, parallel transformation can be re-created in cell culture, employing interrelated criteria for abnormal cell growth. Like the tumor cells in the animal host, transformed cells contain viral DNA encoding the expression of tumor antigens. Finally, all the genetic studies to date demonstrate that the viral tumor antigens are necessary for transformation of cells in culture. They contribute actively to the transformed phenotype (the properties that describe this abnormal cell growth). The activities of these tumor antigens, however, do not explain all the aspects of the transformed phenotype. It is likely that the nature of the host cell—whether mutations in its genome or its expression of specific genes—contributes to the phenomenon of cell transformation in culture.

The Human Papilloma Viruses

Sixty-four distinct human papilloma viruses have been identified and isolated from benign or malignant tumors of the oral cavity, larynx, anogenital region, or skin of human beings—and the list is growing every month. These viruses, close relatives of SV40, also contain double-stranded, circular DNA wrapped around nucleosome subunits made up of cellular histone proteins. The DNA, larger than SV40's (8000 nucleotides instead of 5000), is enclosed in icosahedral particles that are proportionately larger than SV40's but otherwise look very similar. Each of the human papilloma viruses has a slightly different nucleotide sequence. The different types, referred to as HPV-1 through HPV-64, have been linked with specific sets of diseases at particular locations in the body, much like the associations we noted with HSV-1 and HSV-2.

Of the sixty-four types of HPV, a subset is associated with cancer. HPV-5 and HPV-8 may be observed in cutaneous carcinomas in patients with a rare disease called epidermodysplasia verruciformis. These carcinomas usually arise in areas of the body that have been exposed to the sun, suggesting that ultraviolet light acts as a co-carcinogen with HPV-5 or HPV-8 to initiate such cancers. Ultraviolet light can cause mutations in host-cell chromosomes, which might potentially augment the viruses' ability to initiate or contribute to uncontrolled cell growth. This is similar to the host cell's contribution to the transformed phenotype, in that genetic or environmental factors must act along with a virus to cause disease.

About eighteen different HPVs have been isolated from anogenital lesions. These viruses are transmitted by venereal contact. Some HPV types (16, 18, 31, and 33, for example) are termed high-risk because they are found in lesions that can progress to malignant disease. Other types, such as 6 and 11, are termed low-risk, only rarely giving rise to tumors. Certain of the high-risk viruses—HPV-16, HPV-18, or HPV-33—are present in 85 to 90 percent of cervical carcinomas. As is true for all DNA tumor viruses, the HPV chromosome is integrated into the genome of the cancer cells, where a subset of viral genes and proteins is expressed.

A map of the HPV chromosome shows six of the eight early genes (E1 through E8) and two late (structural protein) genes (L1 and L2). In most HPV-related cancers or cancer-derived cell lines, the E6 and E7 genes are actively transcribed, and the resulting proteins can be detected. These observations suggested that the E6 and E7 proteins might be the oncogene products of this virus. This idea gained considerable weight when it was shown that when E6 and E7 were introduced into normal primary cultures of human foreskin cells, these cells were immortalized and transformed for some properties. Mutations in E6 or E7 that cause them to produce faulty proteins result in a failure to transform cells. It is of some interest that the E6 and E7 nucleotide sequences and proteins from the high-risk types—HPV-16, HPV-18, and HPV-33—are distinct from, although closely related to, the nucleotide sequences and proteins of HPV-6 and HPV-11, the low-risk viruses.

About forty years ago, cells from a particular cervical carcinoma were first grown in culture. A derived, immortalized cell line—called the HeLa cell line—became a popular source of human cells, used both for the study of cancer cells and as a medium for the growth of human viruses. At the time, no one realized the association of cervical cancer with HPV; indeed, the human papilloma viruses had not even been isolated. As the role of these viruses in cervical carcinomas became clear, however, it was shown that—after some four decades of growth in culture—HPV-18 genes E6 and E7 were present and were being expressed in the HeLa cells. If the viral oncogenes E6 and E7 were turned off within these cells and their proteins no longer made, the cells stopped growing. After years in cell culture in hundreds of laboratories, the immortality of this transformed cancer-cell line remains dependent upon its viral parasite and the expression of the E6 and E7 proteins.

The Human Adenoviruses

In the days when it was popular for people to have their tonsils and adenoids removed, cells from these organs were often sources of human tissue for study in the laboratory. When cell cultures of

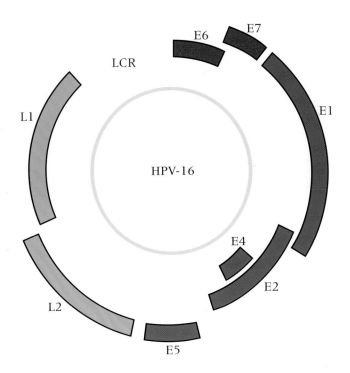

A map of the chromosome of HPV-16, showing the positions of six of the eight early viral genes (E1 through E8) and the two major late viral genes (L1 and L2). The E6 and E7 viral gene products are the tumor antigens expressed in human carcinomas.

human tonsils or adenoids were grown, however, the cells replicated as expected for a while but then mysteriously died. This observation was explained when it was discovered that the cells were being killed by a virus harbored in the adenoids. As it became clear that there was a group of related viruses commonly resident in the respiratory tract, the name adenoviruses was adopted.

The adenoviruses spread rapidly among people living in high-density groups—military recruits in training camps, children in orphanages, and prison inmates. In most cases, these viruses cause mild respiratory distress, conjunctivitis (eye infection), or gastroenteritis (intestinal infection), all readily combated by the body within a few days.

Only in some third-world settings characterized by poor health and nutrition can these viruses lead to more serious disorders.

About forty-two different types of human adenoviruses have been isolated. Because immunity to each virus is effected by a distinct antibody, the viruses are classified by antibody reactivity as types 1, 2, 3, and so on. Each type is associated with a particular disease in the host.

The human adenoviruses fall into six different subgroups, termed A through F, based upon their biological, chemical, immunological, and physical properties. The effects of the adenoviruses seem relatively innocent in humans, but each subgroup contains particular viruses capable of transforming

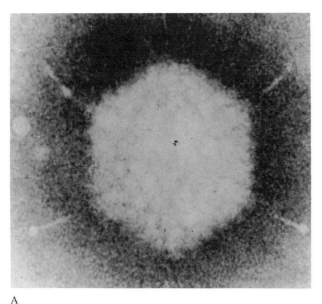

A

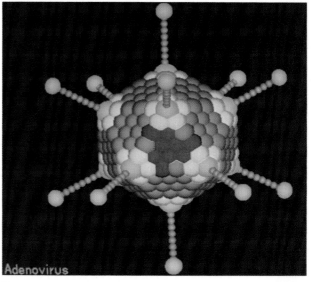

B

A: EM of an adenovirus particle. The hexagonal shape is the icosahedral nucleocapsid, in which the linear, double-stranded DNA is condensed. Fiber projections radiate from the twelve vertices at the fivefold rotational axes. Free capsomeres are also seen, broken off from a virus particle. Like SV40 and the HPVs, the adenoviruses are composed solely of protein and DNA. B: Computer-graphic representation of an adenovirus particle. The icosahedral shell comprises 252 capsomeres, which appear here as blue, yellow, and red subunits. The fibers (green) stick out from the fivefold axes of symmetry.

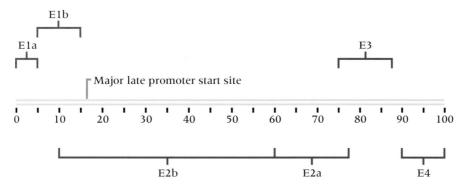

The linear, 36,000-base-pair adenovirus chromosome is represented (in percentages of length), showing the positions of the early viral genes. The E1a genes, expressed first, produce proteins that turn on mRNA synthesis by the other early genes. The E2 genes produce proteins required for viral DNA replication. The midportion of the chromosome contains the late gene promoter, used (after viral DNA replication has begun) to synthesize a very long RNA—30,000 nucleotides—from which some seventeen late proteins are produced in a defined ratio regulated by the rules of RNA splicing. The major late promoter start site is the origin of this long molecule.

animal cells in culture for all six properties discussed in the previous section. Subgroup A is highly tumorigenic: when injected into newborn hamsters, it will induce tumors in 80 to 100 percent of the animals, with a short latent period (3 to 6 months). Subgroup B is moderately tumorigenic: it induces tumors in a lower percentage of hamsters (20 to 60 percent), with longer latent periods (5 to 12 months). Subgroup C fails to induce tumors. Despite their varying effects in living animals, adenovirus agents from subgroups A, B, and C are all able to transform rat or hamster cells in culture with the same efficiency. This curious distinction between the tumorigenic potential of a virus when inoculated into an animal and its ability to transform cells in culture may reflect the role of the immune system in combating some of these tumors. We will return to this point after a brief description of the adenovirus replication cycle.

A protein fiber radiates from each of the twelve vertices of the icosahedral shell surrounding the DNA of an adenovirus virion. It is by means of these fibers that the virion attaches to a susceptible cell. Taken into the cell by endocytosis, the virus is transported to the nucleus; during this process, the uncoating of the virion releases the genetic material, which is a linear, double-stranded DNA molecule about 36,000 nucleotides long. The first viral genes to be transcribed are called the E1a genes; they encode two proteins that stimulate the transcription of the other adenovirus early or delayed-early genes (recall the parallel with the herpesviruses' immediate-early genes).

These delayed-early genes are called the E1b, E2a, E2b, E3, and E4 genes, according to their locations from left to right on the viral chromosome. The E1b proteins facilitate the movement of viral mRNAs into the cytoplasm of the infected cell, at the same time inhibiting the movement of cellular mRNAs out of the nucleus. The E2a and E2b proteins are required for viral DNA replication and include the viral-encoded DNA polymerase. The E3 and E4 genes modulate the host's immune system or other defenses or are involved in regulating viral E1b proteins, as well as cellular gene functions. After these early viral proteins have been synthesized, the viral DNA replicates, producing thousands of copies.

Now late transcription begins, starting at a unique site in the chromosome and proceeding until an RNA molecule 30,000 nucleotides long is completed. This long RNA molecule can then be spliced into five different sets of mRNAs, each of which produces a family of related structural proteins used to construct the virion. Some seventeen late adenovirus proteins are synthesized from all five families of mRNA. The alternative splicing is regulated and creates the late mRNAs in different quantities, so each late adenovirus protein is made in the correct amount relative to the other structural proteins. Finally, the virions are assembled in the cell nucleus, and progeny viruses are released when the dying cells disintegrate.

An examination of which viral-encoded genes are present in hamster tumors or in transformed cells demonstrates that the adenovirus E1a and E1b genes are the viral oncogenes; their products are the tumor antigens. Experiments with mutant E1a and E1b genes demonstrate that one of the two E1a proteins and both E1b gene products are required for the transformation of cells in culture. An E1a gene can immortalize cells by itself, and the E1b gene products confer the loss of contact inhibition and other transformed cell properties. This is an example of two viral oncogenes supplying different parts of the transformed phenotype.

The adenoviruses persist in their human hosts by replicating chronically at low levels in the tonsil or adenoid tissues. This is not a noninfectious latent state like that of a herpesvirus, but a persistent infectious state. In hiding from the immune system, the latent herpesvirus is secure at the price of deferring its chance to replicate for long periods. But the adenoviruses, constantly spreading at low levels, are vulnerable to both the humoral and the cellular arms of the host's immune system.

When killer T cells detect a foreign protein on the surface of a virus-infected cell, they attach their receptors to the foreign antigen and make a specific match. In the virus-infected cell, this antigen is as-sociated with a cellular protein called the class I major histocompatibility antigen; it also must make a match with the similar class I protein on the surface of the T cell. When both sets of proteins are interacting properly, the killer T cell can recognize and destroy the infected cell and limit the spread of the virus. If any of these key elements are missing, however, killer T cells are paralyzed and fail to lyse infected cells.

To avoid elimination by the immune system, the adenoviruses have developed some tricks. At least two adenovirus-encoded proteins have been shown to inhibit the synthesis or insertion of class I major histocompatibility antigens on the surface of infected or transformed cells. One of the viral E3 proteins (E3-19Kd glycoprotein) has been shown to reside in the endoplasmic reticulum (ER) and Golgi membranes of infected cells, where it combines with the class I protein passing through those organelles and prevents the proper addition of carbohydrate to the molecule; as a result, very few class I proteins reach the cell surface. Under these conditions, the T cells no longer identify this cell, failing to recognize that it is infected.

These observations help to explain how the adenovirus persists in the natural host, but they do not account for the widely different abilities of adenovirus type 12 (group A) and adenovirus type 5 (group C) to induce tumors (both viruses transform cells in culture equally well, but only type 12 is tumorigenic). The next set of experiments to be described will shed some light on this issue.

When transformed hamster cells in culture expressing the E1a and E1b genes from adenovirus types 12 and 5, respectively, were inoculated into newborn hamsters (or even adult, immunosuppressed hamsters), all the transformed cells formed animal tumors. When identical sets of transformed cells were inoculated into immunocompetent adult hamsters, on the other hand, those from type 12 made tumors, while those from type 5 did not.

Next, hybrid transformed cell lines were made. One set had type-12 E1a genes and type-5 E1b genes. The second set contained the opposite combination: type-5 E1a genes and type-12 E1b genes. When injected into immunocompetent hamsters, such cells always produced a tumor when the type-12 E1a gene was present—implicating this gene and its protein in the marked tumorigenic ability of this adenovirus type. When cells with the type-5 E1a genes were injected into immunocompetent hamsters, they failed to produce tumors, whether the E1b genes were from adenovirus types 5 or 12.

Finally, the level of class I major histocompatibility antigen was measured in the adenovirus-transformed cells. When the E1a gene came from the highly tumorigenic adenovirus type 12, there was a low level of class I protein in the infected cell's surface, compared to infected cells expressing the adenovirus type-5 E1a gene. Furthermore, transformed cells containing the type-12 E1a gene were resistant to killer T cells, although the T cells could destroy cells transformed by the type-5 E1a gene. One of the functions of the type-12 E1a protein is evidently to reduce the level of the cellular class I major histocompatibility proteins at the infected cell's surface. The E1a protein appears to do this by inhibiting the transcription of these cellular genes.

Clearly, the human adenoviruses have evolved sophisticated defenses to elude the immune system—defenses that have allowed the viruses to persist in their hosts over a human lifetime. In doing so, the viruses have increased their ability to produce tumor cells that can also evade the immune system. There is, however, no evidence that any of these adenoviruses are associated with tumors in human beings. Like SV40, the adenoviruses transform cells in culture, and a subset of them induces tumors in unnatural hosts: rodents. The fundamental mechanism of how viruses contribute to tumor formation—whether in the unnatural host (SV40; adenoviruses) or the natural host (papilloma viruses)—remains the same. One or a few viral-encoded oncogenes either cause the cancer to begin or promote the cancerous growth. These same viral proteins are able to transform cells in culture. It remains a mystery why the adenoviruses and SV40 do not cause cancer in humans.

The Mechanisms of Action of the Viral Oncogenes

Having identified a subset of viral-encoded genes derived from the DNA tumor viruses as oncogenes, our next question becomes: How do the products of these genes function to cause cancer? Answers to this question have begun to emerge over the last two or three years and, remarkably, there appears to be a common theme in the mechanisms used by SV40, the papilloma viruses, and the adenoviruses to induce cancer and transform cells in culture. Furthermore, the way in which the viral-encoded oncogenes function has permitted us to identify cellular genes and proteins that play a critical role in other, nonviral human cancers. Scientists always feel a certain gratification when two apparently separate sets of experiments or questions identify the same gene or protein as a key element in the

The Oncogenes of the DNA Tumor Viruses	
Virus/ oncogene product	Binding site(s) in transformed cells and tumors
SV40	
Large T antigen	Rb, p53
Adenoviruses	
E1a	Rb
E1b-55Kd	p53
HPV-16, HPV-18	
E6	Rb
E7	p53

cause of a human cancer. That is exactly what has happened the last few years in the pursuit of how these viral oncogenes function. To see how this pathway has emerged, we will have to introduce a new set of hypotheses and experiments first conceived by researchers who never even had tumor viruses in mind.

It all began in 1971, when Alfred Knudson of Houston, Texas, was reviewing the epidemiological evidence for the role of inherited gene mutations in the origins of a childhood tumor of the eye. Retinoblastoma, a cancer of the retinoblast (cells that divide to form the retina), occurs exclusively in young children. For some time it had been recognized that this tumor had an inherited component, since 40 percent of patients were from families that had documented past cases of retinoblastoma. The quandary was that in the other 60 percent, retino-

Alfred Knudson.

blastoma was observed in families with no history of this disease.

Knudson, like Denis Burkitt (see Chapter 4), appreciated the value of a single hypothesis that could explain the apparently different origins of these two classes of tumors, the inherited and sporadic forms of retinoblastoma. He pointed out a number of differences between the two forms. Inherited forms all occurred in infants, with a mean age of fourteen months; the tumors always appeared in both eyes; and it was common for more than one independently derived tumor to be observed in a patient (the average was three). Sporadic retinoblastoma (no family history) arose in children already several years of age, occurred only in one eye, and never demonstrated more than a single tumor per patient. It appeared that in families afflicted by the inherited form, the probability of an individual's developing retinoblastoma was about 95 percent, whereas only about one in thirty thousand of the general population was apt to develop a tumor of sporadic form.

Knudson's unified hypothesis explained these two patterns. He proposed that in the inherited form of retinoblastoma, one of the parents passed along a mutated, defective gene—let us call it the retinoblastoma susceptibility gene—so that the fertilized egg had one mutant copy of this gene and one normal copy (from the other parent). During the development of the eye, when retinoblast cells were being produced, a second mutation could occur, within a single cell, in the normal retinoblastoma gene. Now this cell would have two defective copies, and a cancer might develop from it. In sporadic cases of this tumor, the fertilized egg would start with two normal copies of the retinoblastoma susceptibility gene, one inherited from each parent. If a first mutation occurred in this gene during embryological development, and then a second mutation were to affect the corresponding gene on the other chromosome in the same retinoblast cell, a retinoblastoma of the sporadic type would result.

The occurrence of two independent mutations in two identical genes on different chromosomes in the same cell is a very rare event. Its probability is

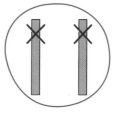

Normal cell
nucleus

Rb mutation in
one chromosome

Rb mutation in
both chromosomes

Knudson's hypothesis to explain the existence of two classes of retinoblastoma in children. The left cell contains two normal Rb genes. The right cell, however, has suffered two mutations, one in each gene, and a cancer occurs because of this. In the inherited form of retinoblastoma, every cell of the body is like the middle cell, containing one mutated Rb gene and one normal gene.

the product of the probabilities of each independent mutation, so it occurs in only one out of thirty thousand people, and then only once per patient. The rarity of this event is also why the sporadic form of retinoblastoma is found in older children; it takes time for both mutations to occur. This contrasts with the relative certainty (95 percent) of obtaining a single mutational event in a normal gene. In most people, this is of no functional significance; the cell's second, normal copy of the gene prevents any negative outcome. But when an individual already has inherited one faulty retinoblastoma susceptibility gene in every cell, cancer occurs on the average three times (three independent mutations) per person, always early in life and almost always in both eyes. This single hypothesis appeared to explain both types of retinoblastoma in the population and, furthermore, to account for the clinical observations.

Inherent in Knudson's hypothesis lay a new concept: that there exists a gene—in this case, the retinoblastoma susceptibility gene—that can contribute to cancer when both copies are defective. This means that the normal function of this gene is somehow to stop abnormal growth, preventing cancer. This class of genes has come to be called tumor-suppressor genes, anti-oncogenes (the opposite of cancer-causing genes), or even recessive oncogenes (because two copies must be altered: the normal gene is dominant over the defective gene).

Over the next fifteen years, the search for the human retinoblastoma susceptibility gene proceeded. First its genetic map location, a position on chromosome number 13, was deduced (we have twenty-three chromosomes, numbered 1 through 23). Finally, in 1986 and 1987, the gene was isolated and identified. Now individuals with inherited retinoblastoma and others with the sporadic form could be tested to determine if the predictions made by the Knudson model were correct. All retinoblastomas tested were shown to have mutations in both copies of this gene. Normal tissue taken from infants with the inherited form of the disease had a mutation in one retinoblastoma susceptibility gene (Rb), while the second copy was normal. Noncancerous tissue from children with the sporadic form had no mutation in either copy of the gene, although their cancer cells had mutations in both copies. All these results are precisely as the hypothesis predicts. An Rb protein, the product of the Rb gene, must exist whose function is to control abnormal cell growth; this protein must be eliminated if a tumor is to develop.

About the time that the Rb gene and protein were being isolated, it became clear that a second gene could also act as a tumor suppressor. This gene encoded a protein called p53, for protein of 53,000 molecular weight (it is common to name a protein by a physical property if the function is not known). In mice that develop a cancer of the cells that produce red blood cells (an erythroleukemia), both copies of the p53 gene must suffer mutations before the cancer can develop. Furthermore, 80 percent of the human colon carcinomas examined to date show a mutation in the p53 gene; no normal copy of this gene is present. Neither in mice nor in humans are these mutations inherited; rather, they arise over the lifetime of the host, accumulating until two mutations are found in the same pair of p53 genes in the same cell. Once it was appreciated that mutations in the p53 gene were prevalent in human colon cancers, a search was undertaken for similar mutations in other cancers. Such p53 mutations now have been observed in a wide variety of human cancers, including carcinomas of the breast, bladder, and lung, as well as astrocytoma (brain cancer), chronic myelogenous leukemia and T-cell lymphomas (blood cancers), osteogenic sarcomas (bone cancers), and others. In fact, p53 mutation appears to be the most common genetic change in the majority of human cancers.

Cancers appear to arise in human beings as a result of a series of mutations that occur in two sets of genes: oncogenes and tumor-suppressor genes. Oncogenes act to promote abnormal growth, and they function even when the cell has a mutation in a single copy of the gene. The other copy may be normal, but the oncogene mutation is dominant and promotes cancer. Tumor-suppressor genes normally act to prevent cell growth. They contribute to cancer by default, when both copies of the gene are lost or damaged; they are recessive in their cancer-causing action (that is, a single normal copy is sufficient to keep cell function normal and noncancerous). Because the mutations that contribute to cancer are rare events that must arise over a lifetime within several genes in the same cell, cancer is by and large a disease of the elderly.

Occasionally a parent may contribute a mutated oncogene or tumor-suppressor gene to his or her offspring, and cancer arises in such a family at a higher-than-expected frequency and earlier in life. A mouse with a faulty p53 gene, for example, gave rise to 106 offspring, 20 percent of which developed cancer. (The expected frequency of cancer in mice with only normal copies of the p53 gene was less than 1 percent over the time period of the experiment.) Recently, human families with high incidences of cancers that most often occur before age 40 have been shown in some cases to inherit a mutant p53 gene, a gene-defect pattern that parallels the case of the inherited form of retinoblastoma. The existence of tumor-suppressor genes and oncogenes has been firmly established as critical in animal and human cancers.

The DNA tumor viruses cause cancer by expressing their specific, viral-encoded oncogene products in cells. For SV40, that product is the large T antigen; for the adenoviruses, the products are the E1a and E1b proteins; and for the human papilloma viruses, the products are the E6 and E7 proteins. What are the functions of these proteins, and how do they act to cause cancer? The cell in the body of an adult host that one of these viruses enters is, frequently, not replicating. Such a resting cell has low levels of the enzymes needed to produce the nucleotide precursors for DNA replication, and it is not synthesizing cellular enzymes, such as DNA polymerase, needed to replicate DNA. In addition, viruses like SV40 and the papilloma viruses require the cellular histone proteins to package their DNA, and these host proteins are only synthesized in actively replicating cells. The viruses, to optimize their own growth and replication, must get the host cell to divide.

These viruses had to find a way to bypass the normal cellular controls over cell division. Two of the critical proteins present in every cell to keep the cell quiescent are the Rb and p53 proteins. The DNA tumor viruses appear to neutralize or alter the functions of Rb and p53 by producing proteins—the viral oncogene products—that bind to and inactivate Rb and p53, or change their functions.

If one examines cells derived from SV40-induced tumors or cells it has transformed in culture, the SV40 large T antigen is found within the nucleus, forming a protein-to-protein complex with the cellular Rb and p53 products. It has also been possible to identify the precise regions of the SV40 large T-antigen protein that bind to Rb and p53. The SV40 T-antigen protein is 708 amino acids long, and each amino acid can be assigned a number—1 through 708—to reflect its linear position in the protein's amino-acid sequence. Amino acids 105 through 114 in this polymer chain are required for binding to Rb, and amino acids 272 through 517 are required for binding to p53. Mutations that change the amino-acid sequences in either of those regions of the protein produce a T antigen that fails to transform cells in culture. Thus, there is a good genetic correlation between T-antigen binding of Rb and p53 and T-antigen function as an oncogene with a capacity to transform cells.

Similarly, in adenovirus-transformed cells, an E1a protein binds to and presumably inactivates Rb protein function, and the E1b-55Kd protein binds to p53. HPV types 16 and 18 produce an E6 protein that forms a complex with p53 and an E7 protein that binds to Rb. The E6 and E7 proteins from HPV-6 and HPV-11, low-risk viruses not often associated with cancers in humans, either bind less well to Rb (E7) or do not bind to p53 at all (E6). The biological property that confers the risk for cancer in the different HPV types correlates, then, with the efficiency of E6 and E7 in binding and altering the p53 and Rb functions within a cell.

Faced with the need to stimulate the replication of the host cell, the DNA tumor viruses have evolved viral oncogenes whose products act upon the cell's anti-oncogenes, which normally function as a brake on the process of cell division. Their protein complexes with Rb and p53 show that these different viruses have targeted common pathways that regulate cell division. Further, these viruses inactivate the very same functions that are lost through mutation in naturally occurring, nonviral cancers. Although the mechanisms of causation in viral-induced and mutation-induced cancers are different, the targets—p53 and Rb—are the same. The DNA tumor viruses, originally studied as good model systems for exploring cancer, have led to the identification of the same genes and proteins recognized by cancer biologists as critical to the origins of human tumors. This is one more example of how fundamental research questions can lead to important understanding of real-world problems.

The Retroviruses

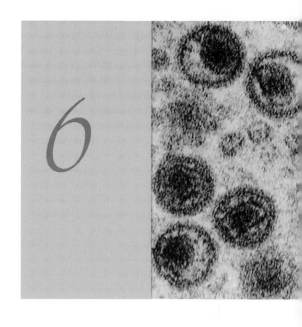

Any living cell carries with it the experience of a billion years of experimentation by its ancestors.

Max Delbrück, 1949

When Ellerman and Bang (1908) and Rous (1911), in their respective experiments with chickens, first isolated viruses that cause cancer in animals (see page 77), they had no idea that they had discovered a remarkable group of agents that some sixty years later would be designated the retroviruses. As these chicken leukemia and sarcoma (solid tumor) viruses were grown and purified, their chemical composition was determined: 60 to 70 percent protein, 20 to 30 percent lipid, 2 percent carbohydrate, and about 1 percent RNA. At that time, these agents were classified as RNA viruses. But in the 1960s, H. Temin and his colleagues made the unexpected observation that DNA replication and DNA-dependent RNA synthesis (transcription of mRNA from a DNA template) were required if a cell infected with viruses of this kind was to produce viral progeny.

To account for this peculiar result, Temin (1964) hypothesized that such an RNA virus, as a part of its life cycle, produces a DNA copy termed a provirus, which in turn is transcribed into viral RNA. This concept was dramatically confirmed when both Temin and D. Baltimore discovered an enzyme in the virus particle itself—the reverse transcriptase—that copied RNA to DNA: the opposite of transcription. This crucial discovery not only introduced precise information explaining how certain viruses replicate but also provided an important new tool that allows molecular biologists to copy any

Left: Newly hatched chicks. Chicken retroviruses were the first viruses to be identified as causes of cancer in animals.

Above: The retrovirus chromosome consists of RNA, not DNA. This EM is of the mouse mammary tumor virus.

RNA species (mRNA) into DNA—a key step in the 1970s revolution of gene cloning and genetic engineering. Lessons from the study of retroviruses, moreover, have formed a cornerstone in our modern understanding of cancer. And one of these agents—the human immunodeficiency virus (HIV)—presents us with a formidable challenge for the 1990s.

Classification, Properties, and Life Cycle of the Retroviruses

Retroviruses acquired their name from events that occur during their replication. Starting with a single plus-strand RNA genome, they copy this into a double strand of DNA. That DNA, in turn, is transcribed, synthesizing an mRNA—identical to the genomic RNA—that is packaged into the progeny virus. All organisms, except for some viruses, store their genetic information in DNA and copy it into RNA (mRNA) to produce proteins; among these exceptions are the retro- (reverse) viruses.

Most of the well-characterized retroviruses infect vertebrates; chicken, mouse, monkey, and human retroviruses have been studied the most thoroughly. Based upon a comparison of the similarities and differences in the nucleotide sequences of distinct retroviruses (these demonstrate the evolutionary relationship or relative closeness of two viruses), seven groups have now been identified and classified.

The first group (the avian leukosis-sarcoma viruses), of which Rous sarcoma virus is the prototype, has been associated with long- or short-latency leukemias or tumors in chickens. Some of these viruses are transmitted between animals (horizontally), while others are inherited through the germ line (transmitted vertically). Viruses that are transmitted vertically are commonly called endogenous viruses; those that are transmitted horizontally are called exogenous viruses. One of the unusual features of retroviruses is that they can, upon rare occasions, enter the germ line of an animal and reside in a host chromosome like a set of normal genes. The viral DNA is then present in the host's sperm or eggs and is passed from parent to offspring. All human beings, for example, carry the evolutionary remnants of retrovirus DNA in our chromosomes as a result of common ancestors who were infected with such a virus. Many varieties of chickens also have an inherited retrovirus.

The second retrovirus group is made up of the mammalian C-type viruses. Their virion structure is a central nucleocapsid surrounded by a lipid envelope. This group is represented by a wide variety of mouse viruses of both the exogenous and endogenous types; the feline leukemia virus, which causes cancer in cats; and simian sarcoma virus, which causes cancer in monkeys. The human genome has integrated several representatives of this virus group, but these endogenous viruses in our DNA are defective: none are able to replicate, because they have sustained mutations that inactivated them over the millennia they have resided in our chromosomes. We recognize these "viral dinosaurs" by their nucleotide sequences, which are closely related to living relatives in this virus group.

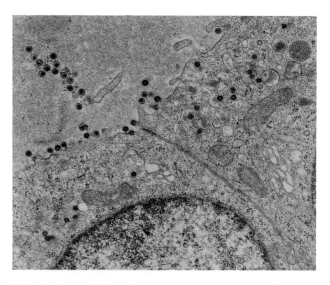

EM of virus particles of mouse mammary tumor virus at the plasma membrane of a cell.

Classification of the Retroviruses

Group	Examples	Properties/pathogenesis
Avian leukosis-sarcoma viruses	Rous sarcoma virus	Carries an oncogene
	Rous-associated viruses (leukosis viruses)	Endogenous and exogenous forms; latter result in B-cell lymphoma
Mammalian C-type viruses	Moloney murine (mouse) leukemia virus	Exogenous; T-cell lymphoma
	Harvey murine sarcoma virus	Carries an oncogene
	AKR murine leukemia virus	Endogenous; T-cell lymphoma
B-type viruses	Mouse mammary tumor virus	Endogenous and exogenous forms; latter transmitted in milk
D-type viruses	Simian AIDS (SAIDS) viruses	Immunodeficiencies
Human T-cell leukemia	HTLV-I, HTLV-II	T-cell lymphoma; neurological disorders
Lentiviruses	HIV-1, HIV-2	AIDS in humans
	Simian, feline immunodeficiency viruses (SIV, FIV)	AIDS-like disease in monkeys and cats
Foamy viruses	Many human and primate viruses	Exogenous; poorly studied

The so-called HERV-C sequences, for instance, represent an ancient, defective, endogenous retrovirus of humans that is present in multiple copies in all humans tested to date.

The third group of retroviruses is the B-type viruses, which have a tightly condensed, acentric nucleocapsid surrounded by a lipid membrane. The major representative of this group is the mouse mammary tumor virus. Because this virus has both an exogenous form and an endogenous form, some types of breast cancers in mice are inherited and some are acquired from virus present in mother's milk.

The fourth group is the D-type viruses, which have virions with condensed nucleocapsids in an acentric position surrounded by a lipid envelope. The surface projections (glycoproteins) in the lipid envelope of D-type retroviruses are less prominent in electron micrographs than are those of the B-type viruses. The SAIDS (simian acquired immune deficiency syndrome) viruses, or monkey AIDS viruses, are in this group.

The fifth virus group is represented by the human T-cell leukemia virus (HTLV-I and HTLV-II) and bovine leukemia virus (BLV). Only exogenous viruses are known in this group, which is responsible for some human or animal cancers and neurological deficits.

The sixth retrovirus group is the lentiviruses (slow viruses), so called because their disease consequences arise only after years of infection. These exogenous viruses are responsible for a number of neurological and immunological diseases but are not directly involved in malignancies. The visna virus of sheep, the equine infectious anemia virus of horses, and the human immunodeficiency virus (HIV) that causes AIDS belong to this group.

The seventh and last family of retroviruses is the foamy viruses; they are called this because they induce extensive vacuolation (vesicle formation) in the cells they infect. Many viruses of this class have been isolated from humans and other primates. These exogenous viruses, not apparently associated with any diseases, have been poorly studied to date and remain uncharacterized.

All retroviruses have a related virion structure composed of similar viral-encoded proteins. Each virion contains two identical RNA molecules, 7000 to 10,000 nucleotides long, which are single plus strands of RNA (that is, the same sense as mRNA). Associated with them is the enzyme reverse transcriptase (RT), whose job is to copy this RNA into double strands of DNA. A second enzyme, called ribonuclease H (RNaseH), is also a part of the RT protein. The viral-genomic RNA is copied into DNA in two stages. First an RNA:DNA duplex hybrid is synthesized. In order to copy the DNA of this hybrid into a double-stranded DNA:DNA duplex, it is necessary to remove the RNA in the RNA:DNA hybrid; RNaseH (H for hybrid) degrades the RNA in an RNA:DNA hybrid molecule, and this is accom-

plished almost as rapidly as the RNA:DNA hybrid is produced.

In the retrovirus virion, a third enzyme is intimately associated with the RNA core. It is called integrase (IN), and its job is to take the double-stranded DNA copy of the virus and integrate—splice—it into the chromosome of the host cell. The IN enzyme is first made as part of the RT protein, but a viral protease cleaves it (cuts it off) from the RT so it can act independently; this viral-encoded protease is the fourth enzyme in the RNA core of the virion. The core is surrounded by the structural proteins of the virus, which form a closed spherical shell about the viral genome.

This nucleocapsid, then, is made up of several proteins, each of which is cut out of a long protein precursor by the viral protease. The capsid proteins are made from the information stored in a viral gene called the gag gene (this gene originally obtained its name from the term group-specific antigen). The gag gene encodes a long polyprotein that has the protease enzyme at one end. Protease cuts itself out of the polyprotein and also cuts the gag protein into several parts, which condense into capsomeres that assemble about the RNA. Similarly, the RT is encoded by the pol gene (pol stands for polymerase), which has at one end the IN en-

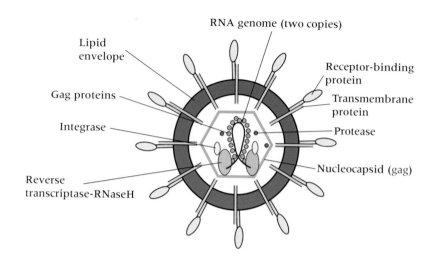

Schematic representation of a retrovirus virion. Two RNA molecules—packaged in close association with reverse transcriptase-RNaseH, integrase, gag proteins, and protease—are surrounded by capsomeres (gag products) to form a nucleocapsid. This is enveloped in a lipid membrane derived from the plasma membrane of the host cell. The env gene products make up the transmembrane subunit and the receptor-binding proteins that project out of the lipid envelope.

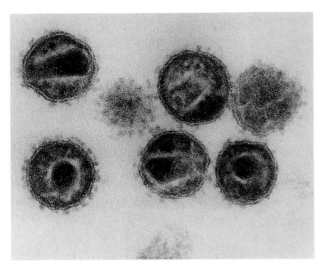

EM of a thin section through a set of retrovirus (HIV-1) particles. The dark nucleocapsid core is surrounded by the lipid envelope (a double layer of dark-light-dark membranes), from which the env gene products project around the virion.

zyme that is cut out of this RT polyprotein by the protease. These proteins and the viral RNA form the nucleocapsid core.

The nucleocapsid core is surrounded by a lipid envelope derived from the plasma membrane of the host cell. Inserted in this membrane is a viral-encoded transmembrane protein, from which projects a protein that attaches the virus to a susceptible host cell. Both of these are glycoproteins, having carbohydrate components that were acquired in the endoplasmic reticulum and Golgi apparatus of the infected host cell. Such a protein is most commonly referred to as gp (glycoprotein), followed by a number that stands for its molecular weight, which is characteristic of each particular retrovirus. These proteins mediate the attachment of the retroviruses to cells and help in penetration of the virions. The env gene (its products project from the lipid envelope) synthesizes a long single protein that is then cut into the two glycoproteins by a protease encoded by the host-cell genome.

All retrovirus genomes contain short nucleotide sequences that are repeated at each end of the viral RNA molecule. The DNA copy of this RNA has a long terminal repeated sequence (LTR) at each end, produced by copying the unique terminal region of the genomic RNA twice. Although the DNA copies of LTRs have several functions, one of the most important is to act as an enhancer-promoter for transcription of genes on the chromosome. A DNA copy of the LTR integrated into the host-cell chromosome attracts cellular transcription factors: proteins that bind to this DNA, recognize its nucleotide sequences, and promote transcription of the viral genes adjacent to the LTR. The order of the viral genes in a retroviral chromosome is always the same: gag-pol-env, bracketed by the LTRs.

In several retroviruses, additional genes and proteins have been inserted into this basic genome plan. In some cases, such as HTLV-I or HIV, these additional proteins regulate the virus—the rate of transcription, splicing of mRNA, transport from the nucleus, or even release of the progeny viruses from cells. These additional genes and functions will be reviewed in the next chapter, on human retroviruses. In some cases, the retrovirus can acquire a cellular gene as part of its chromosome. Occasionally this may result in a tumor-producing virus, like Rous sarcoma virus, and the identification of a cellular oncogene captured by a retrovirus; this phenomenon will be discussed in greater detail in the next section of this chapter.

A retrovirus attaches to a cell by using its receptor glycoprotein to interact with a specific molecule on the cell surface; the best studied example of this, which will be dealt with in Chapter 7, is the adsorption of HIV to specific (CD4) receptors on helper T-cell lymphocytes in humans. This interaction determines in part the host range of a virus (which species or cell type it will infect). Some retroviruses can replicate only in one host species, while others can cross species barriers; the feline leukemia virus of cats, for example, replicates perfectly well in human cells.

After the virion attaches to a cell, the viral membrane and the cellular membrane fuse to-

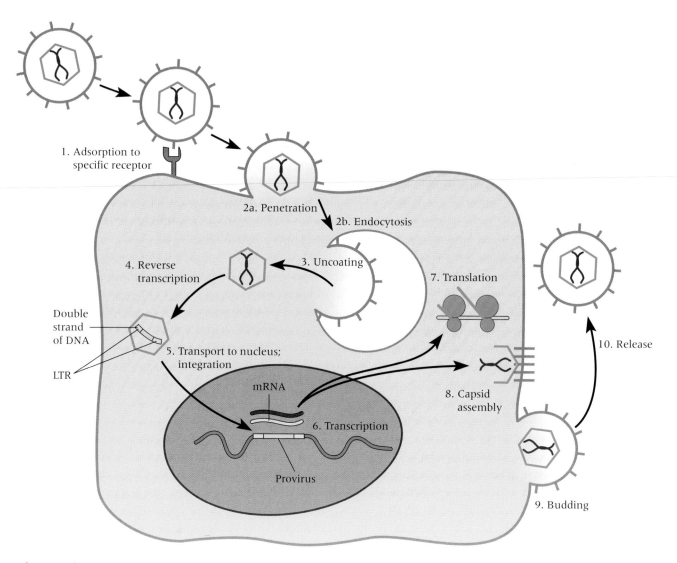

1. Adsorption to specific receptor

2a. Penetration

2b. Endocytosis

3. Uncoating

4. Reverse transcription

Double strand of DNA

LTR

5. Transport to nucleus; integration

mRNA

6. Transcription

Provirus

7. Translation

8. Capsid assembly

9. Budding

10. Release

Schematic drawing of events in the replication of a retrovirus. Once the nucleocapsid core is within the cytoplasm, the reverse transcriptase makes a double-stranded DNA copy of the viral RNA genome, degrading the RNA (RNaseH is used). This double-stranded DNA, still in its nucleocapsid core, moves to the nucleus and is integrated into the host-cell genome (using integrase). The LTR of the viral DNA initiates transcription of the provirus and produces copies of the viral genome (now DNA: the provirus). Some of these mRNAs are spliced and move to the cytoplasm, where they are translated on cellular ribosomes to synthesize gag, pol, and env proteins. The mRNA that is a full-length, plus copy of the viral genome associates with these proteins. At the plasma membrane, budding of virus particles into the extracellular spaces is then initiated.

gether. In the case of HIV, this fusion appears to occur at the cell surface and is mediated by a specific virus function associated with one of the glycoproteins in the membrane. Other retroviruses enter the cell by endocytosis, followed by fusion of the viral envelope and an endosomal membrane. Fusion appears to be triggered by the greater relative acidity of the endosomes, which causes a conformational change in the glycoproteins that promotes membrane fusion and places the virion nucleocapsid and its associated enzymes into the cell cytoplasm.

Between four and eight hours after infection, in the cytoplasm or the nucleus, the RNA in the nucleocapsid is copied (RNA:DNA) and then recopied (DNA:DNA). IN must insert the viral DNA into a cellular chromosome in the specific gene order LTR-gag-pol-env-LTR. By contrast, it appears that the place on a cellular chromosome employed for the integration event—and even *which* cellular chromosome is used—is by and large a random choice, although there is some evidence of a large number of preferred sites. The integrated DNA copy of the viral RNA is now termed a provirus.

The integration event is a required intermediate step in the duplication of all retroviruses. Other plant or animal viruses (hepatitis B, for example) may use reverse transcriptase in their life cycles, but no other animal virus requires an integration event as an intermediate step or provides such a mechanism for stably associating itself with the host. Integration occurs efficiently in every retrovirus-infected cell, as opposed to the rare insertion event that is an aberrant end product for the DNA tumor viruses and that can lead to a tumor in the host. IN recognizes the viral genome at a definite site (the LTRs) and inserts the viral genes in a particular order, as compared to the random order and site observed with the DNA tumor viruses. This is quite a sophisticated evolutionary jump in a virus's or a DNA molecule's ability to reproduce itself.

The net result is that the virus is now inherited in the lineage of the infected cell. Most of the time, retroviruses infect body cells, not the germ line. When they do affect the sperm or egg cells, however, an endogenous virus arises, to be passed on to the offspring of the infected host. Thus, retroviruses are the only viruses with a fossil record—a chemical one.

After integration into a host-cell chromosome, the LTR of the viral DNA promotes transcription of the gag-pol-env genes. The RNAs that are produced are spliced into several different mRNAs, depending upon the retrovirus and the signals it contains. When full-length copies of the LTR-gag-pol-env-LTR are made, they are single plus strands of RNA—the equivalent of the viral genome. The synthesis of these viral genomes is carried out by the host cell's own RNA polymerase. The spliced and unspliced RNAs are transported out of the nucleus onto the cell ribosomes for translation into proteins. The env gene products (glycoproteins) are eventually inserted into the cell's plasma membrane. The RT and other virion enzymes associate with two copies of the viral RNA in the cytoplasm and then move to the plasma membrane, resulting in the budding of particles. Internal cleavage of the gag proteins by protease and condensation of the virus particle complete the retrovirus's maturation—with an envelope derived from the plasma membrane of the infected cell completely surrounding the nucleocapsid core.

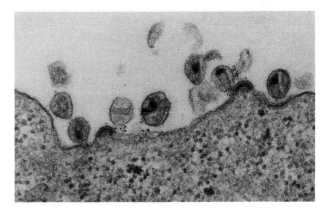

EM of a cross section of an infected CD4 T cell with retrovirus (HIV) particles in the final stages of budding from the cell surface.

This membrane, containing the proteins that attach the virus to its host cell, is essential for viral infectivity. Retroviruses, then, are vulnerable to treatment with detergents (which dissolve lipids in aqueous solutions), alcohol, or other organic solvents. The retroviruses, unlike viruses without lipid envelopes, are generally quite susceptible to the effects of a wide variety of environmental agents. It is not easy to spread these viruses without close contact between hosts.

Some retroviruses never kill their host cells; the infected cells go on to produce large amounts of virus for long periods of time. The progeny of the infected host cell, carrying the provirus in their DNA, will also go on to produce many particles of the retrovirus. Other retroviruses change the properties of the host cell; those that carry oncogenes can transform cells in culture or cause tumors in animals. Yet a third group of retroviruses (including HIV) kill the host cell by mechanisms that are still not well defined. Cancer and neurological and immunological pathologies may result in the host from different kinds of retrovirus infections. In addition, as we have noted, a retrovirus may have an impact on the genetic stability or evolution of a species. We will explore some of these varied and profound effects in the remainder of this chapter and in the next.

The Role of Retroviruses in Long-Latency Leukemias in Chickens

The avian leukosis viruses are retroviruses of chickens, composed of a gag-pol-env RNA gene structure. These viruses infect flocks and replicate in several cell types, including the B cells (antibody-producing lymphocytes). When chicks are first infected with this virus, it has no immediate pathogenic effects; the chicken is healthy, but the virus spreads from B cell to B cell until millions upon millions of infected B cells are producing virus that buds out into the bloodstream. After very long latency times (six months or more), some of these chickens develop B-cell lymphomas. A chronic virus infection is a prerequisite for the development

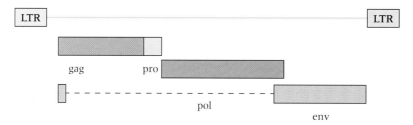

Schematic representation of an ALV provirus and the proteins its genes produce. The LTRs are positioned at either end of the genome. Genes do not literally overlap; rather, one linear sequence yields different mRNAs, depending on where translation begins and ends. So-called reading frames (see page 184) determine the relevant section of the genome. (Compare the letter sequence BATTALION, which yields the words *bat, at, lion, ion,* and *battalion,* among others, depending on where your reading frame begins and ends.) The gag-pro (protease) genes for some retroviruses are read in one continuous reading frame of the genetic code. The pol gene (RT and RNaseH) is encoded by nucleotides in a different reading frame, while the env gene (a split gene with a start in gag) is encoded by the third reading frame. Dashed lines are a convention to indicate splicing.

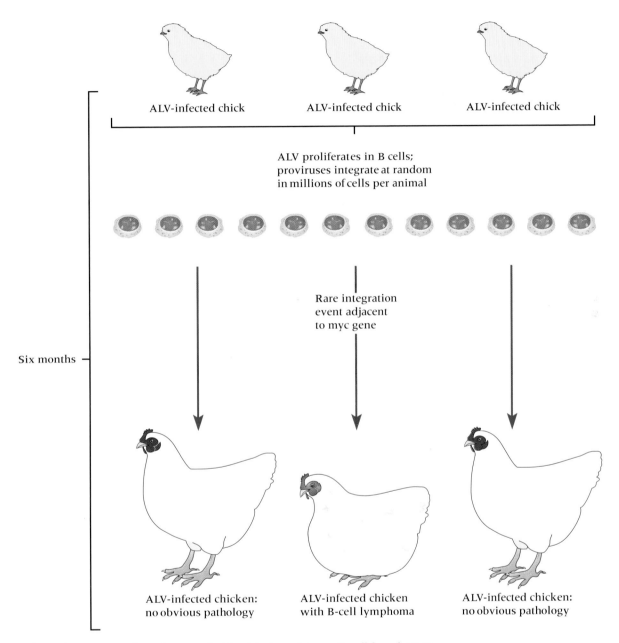

ALV-infected chick   ALV-infected chick   ALV-infected chick

ALV proliferates in B cells;
proviruses integrate at random
in millions of cells per animal

Rare integration
event adjacent
to myc gene

Six months

ALV-infected chicken: ALV-infected chicken   ALV-infected chicken:
no obvious pathology   with B-cell lymphoma   no obvious pathology

Schematic representation of events that lead to long-latency B-cell lymphomas
in chickens infected with ALV. When an ALV provirus integrates adjacent to
the myc gene of a B cell, the viral LTR promotes the transcription of myc
mRNA, increasing the level of myc protein and contributing to abnormal
cell replication. These rare clones are selected for because they increase so
efficiently.

The Retroviruses

of this lymphoma, so something happens over time that initiates a cancer.

It is clear that it is not the virus that changes. If one compares virus taken from chicks with isolates derived from cancerous B cells of older chickens (both normal and malignant B cells produce retroviruses), no difference can be discerned. If, moreover, a retrovirus isolated from a B-cell lymphoma of an older chicken is inoculated into a chick, the latent period to produce a new lymphoma remains the same, about six months; this virus is not better adapted to produce cancers. When, on the other hand, one collects twenty different lymphomas, where each clearly arose as an independent event in a different chicken, all of them have at least one thing in common—the position on the chicken chromosome where the provirus (integrated DNA copy of the virus) resides is very similar in all twenty individual cases.

We noted earlier that retroviruses integrate into the DNA of their host's chromosomes at fairly random positions. That remains true for these avian leukosis viruses. But when millions of B cells are infected with this virus over a long latency period, millions of different integration sites are used. Occasionally an integration event occurring in a gene destroys it—literally cuts it in half. In most cases, there is still another copy of the gene in the homologous chromosome, so the chicken is fine. In other cases, an integration event may kill the cell, but the chicken is not affected by the loss of one or a few cells among millions.

Some integration events, however, occur on a chromosome right next to a chicken gene, in which case the viral LTR, as an enhancer and a promoter of transcription, actually increases the amount of mRNA produced by the adjacent cellular gene. The viral signal (LTR) to make lots of mRNA replaces the normal signal, so that excessive amounts of cellular mRNA are produced. In the twenty lymphomas from twenty different chickens, then, we could look at the gene adjacent to the provirus insertion to see if it is overproduced. When this was done, the insertion of the virus was shown in all cases to be next to or in the chicken gene for myc protein,

which was indeed overproduced in the twenty lymphomas tested.

This is a truly remarkable convergence of observations. As we have seen, in human B-cell lymphomas initiated by EBV infections, two chromosomes fuse—creating a translocation where the break point and spliced joint are adjacent to the human myc gene (the homolog of the chicken myc gene in these B-cell lymphomas). That translocation results in overproduction of the human myc protein. In chicken B-cell lymphoma, the insertion of a retrovirus adjacent to or in the myc gene results in overproduction of the chicken myc protein. In both cases, dozens of different translocations and millions of diverse integration events occur in B cells over long periods of time before the translocation or provirus insertion results in overproduction of myc protein, abnormal cell growth, and malignant replication. These rare events and the occasional cell in which they happen are then selected for, because the cancerous B cell replicates better than do normal B cells. The result is a lymphoma in which all the cells are descendants of this single, rare event, having the same translocation or proviral insertion.

These chromosome rearrangements act upon the same target—the myc gene—and lead to the same result—overproduction of the myc protein and a lymphoma. We do not know how the myc gene functions to promote the growth of B cells, but any cellular gene that is deregulated or altered by mutation and contributes to the development of a cancer is termed an oncogene. The normal form of this gene in humans and chickens, which functions correctly to control B-cell growth, is termed the proto-oncogene. (In cancerous cells, provirus insertion sites on host-cell chromosomes have now been used to identify several adjacent oncogenes in mice and chickens.)

The long latency seen in both these B-cell lymphomas is due in part to the requirement for hundreds to millions of independent events, either translocations or proviral insertions, to take place before one happens at a key spot and is selected for, leading to a cancer. In this case the avian leukosis retrovirus, as a by-product of its replication—which does not in itself harm the chicken—induces a disastrous event by eventually deregulating the critical cellular myc gene.

The Role of Retroviruses in Short-Latency Cancers

Some retroviruses can initiate a cancer in chickens, mice, cats, or other animals very shortly after they are inoculated into a newborn. This occurs so rapidly that it cannot be due to the insertion of a provirus near a critical proto-oncogene. Rous sarcoma virus is an example of this type of retrovirus. Just days after the injection of this virus into the wing web of a chick, a nodule grows into a full-blown solid tumor. As first shown by Peyton Rous eighty years ago, the same virus is produced by these tumor cells, which can be isolated in pure culture, and the disease can be reproduced using this agent—thus fulfilling Koch's postulates.

There is another difference between the long-latency avian leukosis viruses and the short-latency Rous sarcoma virus. If chicken cells in culture are infected with the avian leukosis virus, the virus replicates very well, but the infection has no noticeable effect. Infected and uninfected cells look the same, grow the same, and have the same properties. Infection of chicken cells with Rous sarcoma virus, however, results in their transformation: the infected cells look different (have a different morphology) and grow differently, forming foci in culture dishes and proliferating in agar or suspension cultures. Rous sarcoma virus, then, like SV40 and the adenoviruses, can transform cells in culture. It would appear that the Rous sarcoma virus can do something that its close relative, the avian leukosis virus, cannot.

If we examine the genes of the RNA chromosomes of Rous sarcoma virus (RSV) and avian leukosis virus (ALV), we observe gag-pol-env-X (RSV) and gag-pol-env (ALV). The Rous sarcoma virus contains an extra piece of RNA, provisionally termed X, in its genome, located next to the env gene. The importance of X for the extra biological properties of RSV, compared with ALV, was demonstrated in two ways.

First, certain mutations in X created a virus that could transform cells in culture at 32°C but not at 40°C. This temperature-sensitive mutation, which produces a molecule that functions well at 32°C but folds improperly and functions poorly at elevated temperatures, most commonly occurs in a gene that encodes a protein, as we have seen previously with the SV40 large T-antigen protein. Interestingly, unlike SV40, RSV could replicate itself normally at 40°C, so the transforming function (protein) encoded by the X gene in the virus was not essential for duplication in cell culture.

The second type of mutation in RSV was a deletion of some of the nucleotide sequences in the X gene, resulting in a loss of information (RNA in this case). Like the temperature-sensitive mutation, this one had no effect upon the replication of the retrovirus—but it eliminated the virus's ability to transform cells in culture and to form short-latency tumors in chickens. The X-RNA present in the Rous sarcoma virus seemed to be required for transformation and tumorigenesis but not for retrovirus replication. It was as if an ALV picked up new information in its chromosome, creating an RSV that could now cause cancer in animals.

The next step was to isolate the X gene, now called src because of its association with sarcoma production. This was done during the first half of the 1970s by D. Stehelin, H. Varmus, J. Bishop, and P. Vogt in a series of elegant experiments. They first showed that the src gene is present in RSV but absent in ALV; they then demonstrated that the src gene is also present in normal chicken DNA. Indeed, it is present in the DNA of a wide variety of organisms, including human beings. The virus, during its replicative cycle, picked up a cellular DNA sequence—a proto-oncogene—that, when incorporated in the retrovirus, behaved as an oncogene and acted to cause cancer. What had happened about 1910, or shortly before Rous first discovered this virus, was the incorporation of the src proto-oncogene into an ALV. Peyton Rous, in noticing this tumor, selected the event and isolated the virus for scientific posterity.

This mutation, moreover, could be reproduced close to a hundred years later. When a Rous sarcoma virus carrying the partial deletion of the src gene was put back into chickens, it did not form a tumor for a long time. The virus replicated efficiently with no visible effect or pathology. Insertion of the chicken src proto-oncogene into the virus recurred (very infrequently) via a recombination event between the DNA or RNA of the virus and the host-cell genetic information, thus reproducing a new, fully virulent, Rous sarcoma virus. The occurrence was rare but readily seen as a new tumor in a few of those chickens injected with the deletion-mutant virus. The tumor cells now produced a virus that carried the src gene, and these viruses caused tumors with short latency periods.

A more detailed analysis and comparison of the src oncogene from chicken retroviruses and the src proto-oncogene from the chromosomes of normal chickens revealed that the viral gene had several mutations in its nucleotide sequence, not present in the normal proto-oncogene, that altered its protein function. Furthermore, these mutations were critical in converting the normal gene regulating cell division to a cancer-causing gene whose abnormal product acts to remove control of cell division. Proto-oncogenes, then, are converted to oncogenes by one of two mechanisms: (1) mutations in the gene change critical amino acids in the protein, which then functions abnormally (the src example); or (2) mutations or events conspire to overproduce the normal gene product, which then deregulates cell division (the myc example).

It turns out that src is not the only example of a cellular gene captured in a retrovirus that can cause cancer in animals. To date, researchers have identified twenty-six different cellular genes that can be turned into oncogenes when retroviruses incorporate them into their genomes, initiating tumors. They have also identified oncogenes arising from provirus insertion into a chromosome ad-

jacent to a proto-oncogene like myc. Based on these observations and other experimental approaches, more than fifty oncogenes are now under study.

While retroviruses carrying oncogenes into cells result in cancer in several of their natural animal hosts (chickens, cats, and mice), no retroviruses carrying cell-derived oncogenes have been found in humans. The great majority of human cancers, then, are not due to retrovirus-mediated infections, although we have seen two examples of DNA tumor viruses—EBV and the HPVs—that initiate or contribute to some human cancers (we will review more in future chapters).

This does not mean, however, that oncogenes do not play a central role in initiating or contributing to the origins of human cancers. Clearly, they do; it appears that many human tumors have mutations both in oncogenes and in tumor-suppressor genes, and these mutations accumulate in cells over a lifetime (long latency). When certain combinations of mutant genes are present in specific cell types, they produce a cancer cell that replicates abnormally and results in disease. Not all the oncogenes found in retroviruses appear to contribute to cancers in humans; for example, src mutations have not been found in human cancers, while myc mutations have. Thus, mutations resulting in human cancers appear in only a subset of the oncogenes found in the different retroviruses, but it is to the study of these viruses that we owe the first identification of oncogenes.

A human cell contains forty-six chromosomes (two copies each of twenty-three unique chromosomes) comprising three billion nucleotides in specific sequences that make us what we are. It is estimated that we have 50,000 to 100,000 genes, and therefore the ability to produce that many proteins. The 50,000-plus genes that encode the information for the amino-acid sequences of proteins represent— remarkably—only a few percent of the total three billion nucleotides. A typical gene might be composed of 1000 nucleotides of coding information, producing a protein 330 amino acids long. To identify that gene in the total of three billion nucleotides represents a search for one part per million in the human genome. There would have been no way to find the src gene if the Rous sarcoma virus had not selected it out of the total chicken genome and presented it to Peyton Rous (remember, it functions in human beings, too). The retrovirus not only isolated and incorporated the src gene into its own chromosome, but replicated it, permitted its modification by mutation, and amplified it to give us millions or billions of copies of a single cellular gene, packaged neatly in virus particles. Human cancer is not caused by such viruses, but the retroviruses gave us, in effect, a considerable gift in isolating, one by one, cellular genes that could cause cancer in animals.

The next fundamental problem was to determine the functions of the oncogenes and their proteins, both under normal circumstances and in cancer cells. The workers in this field found themselves in an unusual position. When we study genetics, it is common to start with an observable or functional entity (the phenotype). Gregor Mendel, for example, the first quantitative geneticist, studied pea plants by using leaf color, roughness or smoothness of the pea, and so forth as his observable properties; tracing these phenotypes, he followed the inherited information (the genotype) that conferred these traits through the generations of peas. Ever since Mendel, scientists first had phenotypes and then searched for the genes. The retroviruses, true to their name, reversed this process and gave us the genes without clear functions. The phenotype was the ability to produce a cancer, but the normal function of the gene could not be easily derived from that phenotype. With more than fifty oncogenes presently described, the functions of some proto-oncogene products have begun to be discovered after a wide variety of experimental approaches. Basically, five distinct functions of proto-oncogenes have been identified, each of which appears to be critical in cell regulation.

Gregor Mendel.

The Proto-Oncogenes and Their Functions

To understand what oncogenes do and how their normal counterparts, the proto-oncogenes, control cell division, we will follow the life cycle of a typical cell. The cells of the body remain at rest in the absence of an outside signal. Several genes and proteins, called negative regulators of growth, seem to play a role here, actively preventing cell division. These proteins hold the cell at one of several specific stages of its growth cycle. The stage just before the duplication of DNA is called G_1 (gap-1, to indicate the level of our knowledge), while the DNA-synthetic stage of the cell cycle is termed the S-phase. Two examples of negative regulators of growth are Rb, the retinoblastoma susceptibility gene, and p53; these apparently act as tumor-

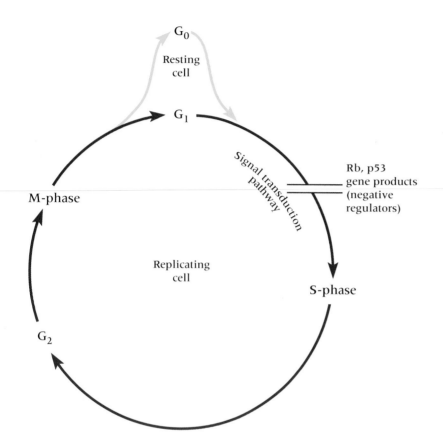

G₀ — labeled as G_0

Resting
cell

G_1

Signal transduction
pathway

Rb, p53
gene products
(negative
regulators)

M-phase

Replicating
cell

S-phase

G_2

Schematic overview of the cell cycle. A resting cell is said to be in the G_0 (gap-0) phase. From outside the cell, it receives a signal to replicate (growth hormone), which binds to a specific receptor on the plasma membrane and is transduced through the membrane by a tyrosine-specific protein kinase that can be a part of the receptor or a separate molecule, like src. Some receptors signal for cell growth by activating a G-protein (ras), which binds the nucleotide GTP. G_1 to S-phase (DNA synthesis) constitutes the signal transduction pathway. In late G_1, proteins encoded for by genes like Rb or p53 act as negative regulators, preventing the cell from entering the S-phase. When this block is overcome, the cell synthesizes its genome. In the G_2 (gap-2) phase, chromosomes condense. M-phase (mitosis) is the stage where the condensed chromosomes separate into the two daughter cells. If the cells are to replicate again, they pass directly to G_1; alternatively, they move to G_0.

suppressor genes, regulating the cell cycle near the G_1-S border. Their gene products are targets of inactivation by DNA tumor viruses (see Chapter 5).

In the process of overcoming this growth control, a hormonal growth factor is secreted and comes into contact with a specific receptor on the surface of a cell. The hormone-receptor interactions are very specific, much like antigen-antibody interactions. A receptor protein that has its extracellular site bound with a hormone sends a signal through the membrane to the inside of the cell for cell division to proceed. The signal may be passed along by chemically modifying a protein; one common modification is the addition of a phosphate group at a specific amino-acid site—a tyrosine. This is carried out by enzymes called tyrosine-specific protein kinases. Some receptors are themselves protein kinases; others attract tyrosine-protein kinases,

which then spread the intracellular signal to divide. The src oncogene is a tyrosine-protein kinase.

Other receptors act by stimulating G-proteins (proteins that bind the nucleotide GTP, guanosine triphosphate) that can amplify the signal to divide. G-proteins exist in either of two states—one signals for cells to divide, the other does not. To prompt cell division, a G-protein binds to the nucleotide GTP; a receptor bound to its growth hormone will promote the GTP form of the "switch": the cells respond by dividing. But when the GTP is converted to GDP (guanosine diphosphate) by the removal of a phosphate group (using an enzyme termed a phosphatase), the protein no longer gives a positive signal for cell division.

Both the G-proteins and the tyrosine kinases appear to transmit their signals for cell division by means of proteins that regulate mRNA production from selected genes in the cell nucleus by binding to enhancer-promoter signals adjacent to a gene. These proteins are called transcriptional regulators; myc is an example of this group. The cellular genes regulated by transcription factors of this type are the ones needed to commit the cell from G_1 to S-phase, synthesize nucleotides, replicate DNA, and duplicate the genetic information. When this process is completed, the cell enters a second gap phase (G_2) to prepare for the division of chromo-

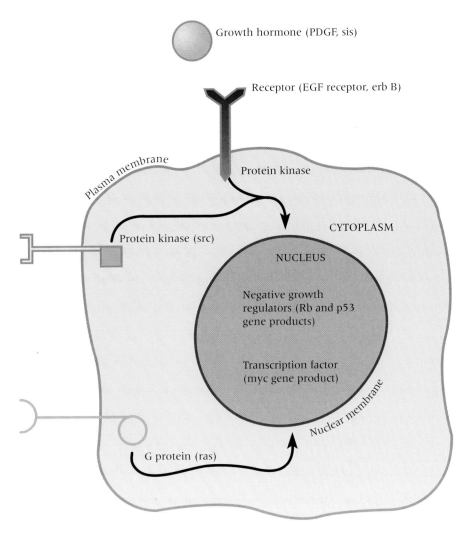

Growth hormone (PDGF, sis)

Receptor (EGF receptor, erb B)

Plasma membrane

Protein kinase

Protein kinase (src)

CYTOPLASM

NUCLEUS

Negative growth regulators (Rb and p53 gene products)

Transcription factor (myc gene product)

Nuclear membrane

G protein (ras)

Schematic diagram of a cell's growth regulators—growth hormones; receptors; protein kinases; G-proteins; and transcription factors. (Sample proteins are given in parentheses. Growth hormones and their receptors, however, are matched pairs; PDGF, for example, will only bind to and stimulate its own specific receptor). These are the protein-mediated pathways of growth signal transduction. All are products of one of the five classes of proto-oncogenes, which, when mutated into oncogenes, act to drive cell replication abnormally.

Selected Examples of Oncogenes Acquired by Retroviruses

Retrovirus	Oncogene	Proto-oncogene (cellular homolog) function	Class of proto-oncogene products (signal transduction pathway factors)
Simian sarcoma virus	sis	Platelet-derived growth factor (PDGF)	Growth factor
Avian erythroblastosis virus	erb B	Epidermal growth factor (EDF) receptor	Growth factor receptor (tyrosine kinase)
Murine sarcoma virus	ras	Unknown (growth signal)	G-protein (receptor signal)
Rous sarcoma virus	src	Unknown (growth signal)	Tyrosine kinase (receptor-associated)
Avian myelocytoma virus	myc	Regulates gene expression	Transcription factor (nuclear)

somes and cytoplasm into two cells. This is accomplished in the M-phase (mitosis) of the cell cycle, in which each chromosome is partitioned into two daughter cells. Each of these steps is regulated by proteins from genes that control the cell cycle (G_1-S-G_2-M) and are expressed at one or a few stages in this sequence.

Each of the critical genes and functions in this cell cycle may be a proto-oncogene. The sis oncogene, for example, was first isolated as part of the RNA chromosome of the simian sarcoma virus, producing tumors in monkeys; its product is a mutated form of a growth hormone called platelet-derived growth factor (PDGF). When we are cut, a clot forms to prevent the loss of blood. Chief among the cells that promote this clot are blood cells called platelets: nets of fibrin proteins on their surfaces plug the wound and stop the bleeding. At the site of injury, platelets release PDGF, which acts upon the cells near the wound (the fibroblasts) and stimulates their replication to repair the injury. PDGF is, therefore, a hormonal growth factor and, when altered, an oncogene.

The erb B oncogene, isolated first from the avian erythroblastosis virus, causes cancer of the cells that are precursors of red blood cells. The proto-oncogene of erb B—that is, the normal copy of this gene in people—encodes a cell-surface receptor of epithelial cells. In the same wound that PDGF acts on, an epidermal growth factor (EGF) targeting the epithelial cells of the skin is also produced. The erb B proto-ongene is the receptor for EGF.

The ras oncogenes that are isolated from sarcoma viruses are G-proteins. Even in the absence of the nucleotide GTP, these proteins signal for continued cell growth. As we saw previously, the oncogene from the Rous sarcoma virus, src, is a protein kinase that adds a phosphate group (a signal for growth under some conditions) to a tyrosine site in a receptor protein that has not yet been identified.

Finally, the myc oncogene is normally a transcription factor that recognizes a specific nucleotide sequence in the genome regulating several critical genes involved in cell division.

The adjacent table lists some of the oncogenes captured by retroviruses and their proto-oncogene functions in the normal cell. It is clear from this survey that each step in the signal transduction pathway regulating cell division can potentially create an oncogene when a normal gene is altered.

In human cancers a subset of these oncogenes, such as ras, has been found to be mutated in many malignant cells. In human chronic myelogenous leukemia, for example, a tyrosine kinase called abl (first found in the Abelson murine leukemia virus) commonly occurs in an altered form because of a chromosome translocation. This disease has a preliminary phase that lasts several years, with all the cancer cells containing an abl translocation. A few rare cancer cells then undergo a further mutation in the p53 tumor-suppressor gene, and this (possibly with other events) results in a rapid growth of cells that is usually the terminal phase of this cancer.

It appears that most human cancers arise over a lifetime by the continuous introduction of mutations in proto-oncogenes and tumor-suppressor genes. When a single cell has accumulated several of these mutations, it loses its growth control and divides abnormally, and a cancer ensues. This is why most cancers are found in older individuals; it takes a long time to accumulate five or six mutations in the same cell. As we saw in reviewing Knudson's work on retinoblastoma (Chapter 5), this hypothesis also explains why some cancers seem to occur in families—often, in those cases, at earlier ages. If someone inherits one of these genetic alterations, the mutation is present in all cells of the body, and one less mutation has to occur over a lifetime to initiate a cancer. Both Rb and p53 genes can have inherited mutated forms, and a high incidence of cancer in a family is frequently the result. Although most human cancers, as we now understand them, do not involve retroviruses or even DNA tumor viruses, the study of these viruses and how they cause cancer in animals not only has led us to the oncogenes and tumor-suppressor genes, but also has reinforced our concept of the unity of genes, gene products, and the mechanisms regulating life processes in all living things.

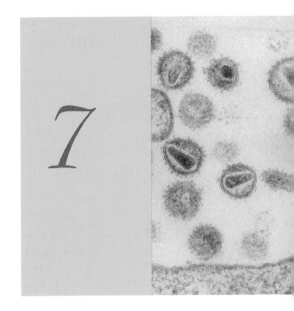

The Human Retroviruses

7

Infectious disease is one of the few genuine adventures left in the world. The dragons are all dead and the lance grows rusty in the chimney corner. . . . About the only sporting proposition that remains unimpaired by the relentless domestication of a once free-living human species is the war against those ferocious little fellow creatures, which lurk in the dark corners and stalk us in the bodies of rats, mice, and all kinds of domestic animals; which fly and crawl with the insects, and waylay us in our food and drink and even in our love.

Hans Zinsser, 1935

Once the retroviruses were shown to play a central role in many naturally occurring animal cancers, research was organized to identify human retroviruses that might cause cancer. No human cancers appeared to have an infectious or epidemic origin, but the animal models of retrovirus diseases and our growing knowledge of oncogenes made it worth a serious attempt to look for such agents. Initial results were disappointing. By the last half of the 1970s, several claims for the existence of human retroviruses were found, upon closer inspection, to be based on contaminating retroviruses of animal origin. Then several new developments turned the search in a more productive direction.

Left: Acquired immune deficiency syndrome (AIDS), identified only a decade ago and traced to certain human retroviruses, has taken a devastating toll of lives around the world. The NAMES Project AIDS memorial quilt commemorates some of the victims in its thousands of panels.

Above: EM of the human immunodeficiency virus (HIV).

Human T-cell Leukemia Virus I (HTLV-I)

In 1977, K. Takatsuki and his colleagues realized that the pathology of a variable T-cell leukemia in Japanese adults was characterized by a unique set of properties and could be described as a single disease syndrome. Adult T-cell leukemia (ATL) had formerly been classified as several different disease entities, leading to a confusing set of diagnoses, treatments, and epidemiological studies. As was the case for Burkitt's lymphoma, this newly unified syndrome was rapidly shown to have a peculiar geographical distribution. In Japan, ATL was found predominantly in the southern islands (Kyushu and Okinawa) and the northernmost island (Hokkaido). The largest island of Japan (Honshu) showed only sporadic cases of ATL, mostly in isolated coastal villages. Such geographic patterns suggest that genetic, environmental, or infectious agents are interacting to produce a disease.

Meanwhile, Robert Gallo and his colleagues at the National Institutes of Health (NIH) in Bethesda, Maryland, had been in the thick of the search for a human retrovirus and were examining virus particles produced by a human T-cell lymphoma in cell culture. The virus, which had a reverse transcriptase activity, was an immunologically distinct entity that they called human T-cell leukemia virus I (HTLV-I)—the first clear-cut example of a human retrovirus to be reported. The next year (1981), I. Miyoshi and Y. Hinuma announced that a cell line derived from adult T-cell leukemia patients contained retrovirus particles that they had visualized in the electron microscope. Furthermore, Hinuma and his colleagues showed that patients with ATL produced antibodies that detected antigens in this retrovirus-infected T-cell line. Only the T-cell line harboring the retrovirus reacted with these antibodies, which were not present in most Japanese. Hinuma called his retrovirus adult T-cell leukemia virus 1 (ATLV-1).

A specific human leukemia had been identified; a specific human retrovirus had been isolated.

Antibodies from the leukemia patients reacted against the retrovirus, cultured in a cell line from a patient with ATL. HTLV-I and ATLV-1 were eventually shown to be the same agent; because the initial isolation was called HTLV-I, that name was retained for the virus.

Over the next few years, the evidence linking HTLV-I to ATL was considerably strengthened. First, the unusual geographical distribution of ATL in Japan was shown to be identical with that of HTLV-I. Further, new pockets of HTLV-I were discovered in the coastal regions of central Africa and,

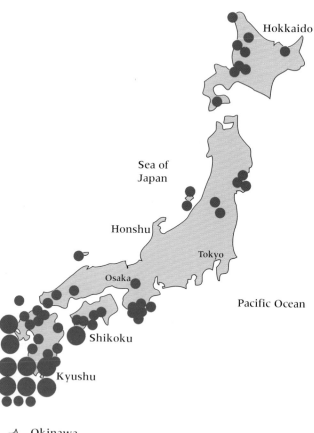

Map of Japan showing distribution of adult T-cell leukemia. The larger the dots, the more cases.

less frequently, in the Caribbean basin, in Taiwan, and among the aborigines of Papua New Guinea; in all these cases, ATL was associated with HTLV-I. The HTLV-I isolates from these diverse geographical locations were, moreover, very homogeneous—some 96 to 99 percent of the nucleotide sequences in the viral RNA were identical. Second, it was shown that virtually every ATL patient tested was infected with HTLV-I. Third, tumor cells taken from ATL patients could be grown in culture (immortalized), and all contained at least portions of an HTLV-I provirus in their DNA. Fourth, when HTLV-I was incubated with normal human T cells, the viral DNA could integrate into the chromosome of these cells, express its viral proteins, and immortalize the cells for growth in culture. Fifth, two animal retroviruses (bovine leukemia virus and simian T-cell leukemia virus 1) were shown to be relatives of HTLV-I, by demonstrations that the sequence of nucleotides and the organization of viral genes in the RNA chromosomes of all three are similar. It was possible to fulfill all of Koch's postulates with each of the animal agents, proving that the respective retroviruses cause bovine or simian leukemia. Medical ethics prohibit definitive experiments with human beings, but it can be inferred from the strong epidemiological evidence that HTLV-I causes ATL. But it remains a mystery how HTLV-I infects human beings over a twenty- to forty-year latency period and then promotes a T-cell leukemia in only about one-tenth of one percent of those infected.

One possible clue came from a detailed study of the way the virus is transmitted from person to person. Despite its peculiar localization, HTLV-I is not endogenous (genetically inherited). Rather, it is transmitted from mother to offspring in utero—via passage of infected maternal lymphocytes across the placenta—or during nursing, through infected T cells in the mother's milk; thus, it may be acquired quite early in life. HTLV-I is also transmitted during sexual intercourse by means of infected lymphocytes in the semen. A third route of transmission is through blood or blood products that contain whole white blood cells. In virtually all cases, HTLV-I must be transmitted via an infected cell, not as a free virus. It appears that HTLV-I is

very poorly infectious—multiple exposures via breastfeeding or sexual intercourse are required to pass it from host to host. As we shall see, this inefficient transmission is the reason for the unusual geographic distribution.

Once we recognize how HTLV-I is transmitted, we can understand why this virus is prevalent in certain high-risk groups. Intravenous drug abusers who pass white blood cells back and forth in shared needles have a high incidence of this virus. Nine percent of the sexual partners of such drug abusers have HTLV-I antibodies, demonstrating exposure. In Trinidad, 15 percent of male homosexuals have antibodies directed against HTLV-I (as opposed to 2.4 percent of the general population), indicative of the passage of the virus through semen during anal intercourse. Multiple sexual partners also help spread the virus.

An indication of how poorly HTLV-I is transmitted comes from comparing the incidence of antibodies to HTLV-I, a virus endemic to Trinidad, with that of antibodies to HIV, a virus only recently introduced there. As noted, 2.4 percent of the total Trinidadian population have HTLV-I antibodies, while 1 percent have HIV antibodies. In the male homosexual community, 15 percent have HTLV-I antibodies, as contrasted with the more infectious HIV, to which 40 percent of these individuals have developed antibodies.

Despite chronic, early-in-life exposure to HTLV-I, only one in one thousand to one in ten thousand virus carriers over the age of forty develops this disease. Based upon a study of the presence of antibodies directed against HTLV-I in the Japanese population, there are presently about one million infected carriers there in a total population of one hundred and ten million. The incidence of ATL in Japan is about three hundred to five hundred new patients per year.

As might be suggested by the geographical distribution of ATL and HTLV-I in Japan, the ethnic background of those affected by this virus is not homogeneous. There are at least three ethnically distinct populations in Japan: the Ainu, residing predominantly in the north (Hokkaido); the Ryukyuans, mostly located in the south (Okinawa); and the Wajins, the great majority of modern-day Japanese, largely clustered in the center of the country. The Ainu and the Ryukyuans appear to share some genetic and physical traits; their ancestors are considered to have inhabited Japan since 10,000 B.C. The Ainu, in particular, are the descendants of a preagricultural society in the Jomon period, more than twenty-three hundred years ago. By contrast the Wajins, arriving in Japan between 300 B.C. and A.D. 600 from the mainland of Asia, are genetically distinct from the isolated populations of Ainu and Ryukyuans. When, in the 1980s, the frequency of antibodies directed against HTLV-I was examined in the three groups, the results—Ainu, about 45 percent antibody-positive; Ryukyuans, about 30 percent antibody-positive; and Wajins, about 1 percent antibody-positive—reflected these genetic and ethnic differences.

For social and cultural reasons, as well as because of the isolated coastal and rural life style of the Ainu and Ryukyuans, these ethnic groups do not generally interact with each other socially and have remained genetically distinct. This suggests that the indigenous population of Japan before 300 B.C. harbored the HTLV-I virus, while the newer immigrants, the Wajins, arrived uninfected from mainland Asia. (Modern Chinese do not have HTLV-I antibodies, and no carriers are found in South Korea, where the Wajins originated.) Because the virus passes from parent to offspring in mother's milk and from adult to adult via sexual intercourse, isolated ethnic groups tend to pass it on between individuals (horizontally) almost as if it were genetically (vertically) transmitted.

How does this set of observations explain the origin of ATL? How does the virus associated with adult T-cell leukemia persist in a host group, especially when that virus is transmitted so poorly? The answers to these questions are still speculative, but the best hypothesis has been put forth by Yorio Hinuma, who first found the retrovirus particles in T-cell cultures from ATL patients. He points out that between forty thousand and one hundred thousand years ago, a major migration of human beings occurred into different continents, where—

joining some existing populations—the Europoid, Negroid, Australoid, and Mongoloid lineages developed in geographic (and genetic) isolation from each other. Hinuma theorizes that these migrants brought with them a high incidence of HTLV-I carriers. But because the twenty- to forty-year latent period before the onset of ATL was longer than the average human life span fifty thousand years ago, HTLV-I infection had very little effect upon the longevity and breeding capabilities of humans. Each of the migrating groups then gradually lost its virus carriers because the efficiency of transmission from husband to wife or mother to child is lower than 50 percent. Only groups that were genetically and culturally isolated, with repeated intermarriage from generation to generation, could maintain carriers. The HTLV-I carriers that now remain in the world are the descendants of these isolated peoples, such as the Ainu and Ryukyuans of Japan, some black populations of Africa (and their descendants in the Caribbean basin), and the aborigines in Papua New Guinea.

There are several ironies here that point out how change and modern technology affect the evolution of a virus. Once again, we see how advances in health care have, by prolonging our life span, introduced a disease—ATL—that did not exist before. Furthermore, HTLV-I, destined to be lost from its host population because of poor transmission rates, has found a renewed life in intravenous drug abusers and through multipartner sexual practices that spread it more efficiently. The use of whole-blood transfusions in the last half of the twentieth century, moreover, provided this virus with an independent route of transmission and access to previously unreachable hosts. Contemporary technologies and modes of behavior have changed the environment of this retrovirus and given it a new lease on life just when it was on the verge of extinction.

With the number of HTLV-I carriers increasing and the incidence of ATL on the rise, a critical question remains to be explored: How does this virus cause cancer? HTLV-I is a typical retrovirus, with two copies of an RNA genome, a little over 9000

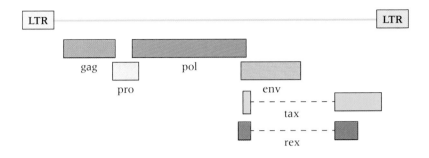

Schematic map of the HTLV-I chromosome. The DNA provirus has terminal repeated nucleotide sequences (LTRs) bracketing the gag, pro (protease), pol (RT), and env genes. The gag and pol genes are in a different reading frame (see page 184) from the pro and env genes. Two additional genes, tax and rex, regulate viral and cellular gene expression. Dashed lines are a convention for mRNA splicing.

nucleotides long, encoding for gag-pol-env genes surrounded by short terminal repeat sequences (see the figure above). The gag gene produces three structural proteins that form the core of the virus surrounding the RNA. Each of these proteins is cut out from a larger precursor protein by the viral-encoded protease (pro), whose gene resides between and overlaps with the gag and pol genes (sequences of nucleotides). The env gene encodes a gp46 and a p21 protein, which make up the spikes that protrude from the lipid envelope surrounding the ribonucleoprotein core of the virus. There is a region of the viral RNA chromosome about 1700 nucleotides long, located between the env gene and the long terminal repeat of the DNA provirus, that encodes the information for two proteins. Because this structure looks a lot like the avian retrovirus that encodes the src oncogene between the env gene and the proviral DNA LTR, a first thought was that HTLV-I might have acquired a cellular gene that initiates ATL. A test of this possibility quickly eliminated the idea; the 1700 nucleotides of this HTLV-I region had no homologous nucleotide sequences in the human genome. HTLV-I did not carry an oncogene that derived from the host.

A second hypothesis—that HTLV-I infection over a lifetime resulted in a rare integration of the provirus into a specific location in a human chromosome, leading to an increased level of expression of an adjacent human gene, which in turn promoted a cancer—was also tested. While many of the leukemic T cells taken from any given individual patient had a single provirus integrated in the same site or chromosome (showing that ATL-containing cells derive from a single, clonal event), when the integration sites of these proviruses from many different ATL patients were compared, no common site was found. Different people with ATL have different integration sites. The avian leukosis virus B-cell lymphomas and Burkitt's lymphoma all activate the myc gene, but this mechanism does not seem to apply to ATL.

Because one region of the HTLV-I provirus chromosome can encode a set of two genes (the 1700 nucleotides between env and LTR)—which are not observed in the avian or mouse retroviruses—attention began to focus on these gene functions. In a series of experiments, it was shown that these genes produce two viral proteins, called tax and rex, that regulate the levels of viral mRNAs and proteins in infected cells. Both tax and rex appear to be required for efficient virus replication, although neither is found in the virus particle. Tax, acting upon the LTR adjacent to gag, vigorously promotes the transcription of new mRNAs, producing more gag-pol-env transcripts and more viral RNA chromosomes; the tax protein is, therefore, a transcriptional activator protein. Rex appears to regulate the splicing and processing of the viral mRNA. Acting at the mRNA level after large numbers of transcripts are made from the proviral DNA, rex seems to favor the production and transport of gag, pol, and env mRNAs into the cytoplasm of the infected cell and to minimize the amount of tax and rex made. Rex promotes more viral structural proteins in an infected cell.

What has all this to do with producing a cancer cell? T cells, like many cells, are stimulated to grow by the presence of an extracellular growth factor, which in this case is called interleukin-2 (Il-2). This protein acts upon the Il-2 receptor in the plasma membrane of T cells. T cells are usually kept quiet and nondividing because the levels of Il-2 and Il-2 receptor are strictly regulated. Exposure to a foreign antigen triggers the signals to synthesize Il-2 and the Il-2 receptor, both of which are required to initiate T-cell division to eliminate the foreign invader. Remarkably, tax not only promotes the transcription of the LTR-gag-pol-env genes but also stimulates the transcription of the cellular genes for Il-2 and the Il-2 receptor; that is, tax acts powerfully upon the enhancers of cellular genes to promote T-cell division. In fact, tax appears to stimulate the transcription of a number of cellular genes, including two cellular proto-oncogenes, fos and platelet-derived growth factor (PDGF, the proto-oncogene of the sis oncogene). Fos is known to promote the transcription of some cellular genes on the pathway to cell growth and division. Tax, in addition to stimulating growth factors and cell receptors, seems to begin a cascade of events in which many cellular genes are activated to promote growth. Tax appears to be a viral-encoded oncogene, inappropriately turning on cellular genes and proteins. The possible role of rex in this process is less clear.

At this juncture, the cautious reader should not be entirely satisfied with this answer to the question: How does HTLV-I cause ATL? The presence early in life of HTLV-I, producing tax, which promotes cell division, hardly explains why it takes some twenty to forty years to stimulate a clonal event in a small number of infected individuals that results in ATL. Furthermore, patients with ATL induced by HTLV-I have few leukemic cells expressing any viral proteins, suggesting a role for the virus early in the development of the cancer but not in its maintenance or later stages. Clearly, tax does not explain all the facts; it is likely that the immune system and its ability to recognize infected T cells play a role here, as well.

When HTLV-I is added to a culture of resting (nondividing) T cells in a culture dish, transformed and immortalized cells proliferate over the next four weeks, and clones of cells can be generated fairly rapidly. The cells frequently produce Il-2, and the Il-2 receptor is expressed on their surfaces. All this happens rapidly in cell culture when compared to the same events going on in a human being. We need to understand just what happens over the two- to four-decade period of chronic HTLV-I infection in humans and why only one in one thousand to one in ten thousand HTLV-I carriers develops ATL. The answers may provide the foundation for understanding basic disease processes in human beings.

Human T-cell Leukemia Virus II (HTLV-II)

HTLV-II was first isolated in 1981 from a T-cell line in culture (called Mo-T) that was derived from the spleen tissue of a patient (Mo) with hairy-cell leukemia. (This rare leukemia is so named because the cells put out long surface processes resembling hairy projections.) HTLV-II is related to HTLV-I immunologically (some antibodies show that these two viruses are distinct, but other antibodies react with both viruses, showing a relationship); and, as we shall see, they share some nucleotide sequences. In 1985, a second patient infected with HTLV-II was diagnosed with a T-cell hairy-cell leukemia with some similarities to that of the first patient. HTLV-II was present as an integrated provirus, and leukemic cells resulted from a clonal proliferation. Unfortunately, these isolates of HTLV-II, plus two additional reports with less detailed information, are too few incidences for us to conclude that the virus causes this disease.

Furthermore, while most hairy-cell leukemias are B-cell leukemias that are HTLV-II negative, even rare T-cell hairy-cell leukemias with no de-

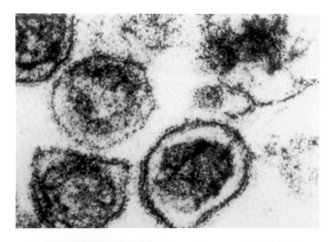

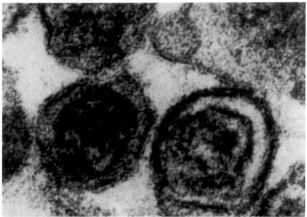

EMs of the virions of HTLV-I and HTLV-II.

The genetic structure of the HTLV-II genome has been studied; it shares 65 percent of its nucleotide sequences with HTLV-I (remember, HTLV-I isolates are 96 to 99 percent identical in their nucleotide sequences). Like HTLV-I, the HTLV-II agent has an RNA genome with gag-pol-env genes and both the tax and rex genes organized in an identical fashion. Clearly, these viruses are close cousins in an old family of viruses that have evolved with us over the past hundreds of thousands of years. This is in stark contrast to the third example of a human retrovirus, whose origins remain unclear but appear to be much more recent and whose pattern of disease was thrust upon us as the major epidemic of the past ten years: the human immunodeficiency virus.

The Human Immunodeficiency Virus (HIV)

By the last two decades of the twentieth century, clearly the century of science and its triumphs, we had come to feel a certain security derived from our insights and new technologies. Progress in the health sciences, represented by the discovery of antibiotics, the introduction of new vaccines, and the development of complex treatments for diseases that had long been lethal, had given us long and productive lives. A serious threat to both our lives and our perceptions became apparent in 1981. The story of the realization that a new disease—AIDS—had come upon us and of the hunt for its cause is a study of people, science, and society: a story all the more important because we will continue to live with its consequences in years to come. It is a story with many lessons.

Recognition of the AIDS Syndrome

Over many years, local, state, and federal governments in the United States have come to appreciate the need for good medical surveillance of the popu-

tectable HTLV-II have been reported. For this reason, the term atypical hairy-cell leukemia has been used for HTLV-II associated cases. At least two additional HTLV-II cases have been seen without apparent evidence of malignancy, one in an AIDS patient (immunosuppression may have permitted this virus to replicate) and the other in a hemophiliac who had had multiple blood transfusions. A recent study of intravenous drug abusers in New Orleans showed a surprisingly high incidence of HTLV-II in their lymphocytes. Clearly, a real understanding of the role of this virus in human disease must await further isolates and information.

lation. At the federal level, this is done by the scientists at the U.S. Centers for Disease Control (CDC), located in Atlanta, Georgia. One of the CDC's functions is to dispense medical supplies that are categorized as experimental because they are used to treat rare diseases. The major pharmaceutical companies do not usually develop drugs for very uncommon diseases; when only one or two patients a year need such an expensive drug, clinical trials to test its efficacy are not practical. When a drug may be effective but has not been tested extensively, it is classified as experimental and dispensed by the CDC, which keeps good records on the incidence of rare diseases in the United States.

It was just this procedure that proved telling between September 1980 and May 1981. Over that eight-month period, Sandra Ford, a technician responsible for rare drug orders at the CDC, had five requests for the drug pentamidine isethionate to treat an unusual type of pneumonia caused by the protozoan *Pneumocystis carinii*, which lives with humans but rarely causes disease. Most *Pneumocystis* pneumonia cases occur in individuals undergoing immunosuppressive chemotherapy for cancer or in transplant patients using drugs to block immune-system function and potential organ rejection. Under these conditions, with no immune cells to fight off opportunistic infections, normally harmless organisms can cause disease.

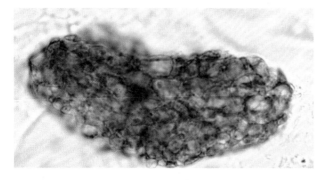

Light micrograph of *Pneumocystis carinii*.

From November 1967 to December 1979, the CDC had had only two requests for pentamidine to treat patients who were not taking immunosuppressive drugs; the occurrence of five requests in an eight-month period was decidedly strange. When Ford brought this to the attention of her supervisor, a group of epidemiologists examining these data noted that all five cases were in Los Angeles and involved young homosexual men whose immune status would have been expected to be normal. In June 1981 the weekly CDC newsletter published a note about these five cases of pneumonia caused by the *Pneumocystis carinii* organism in previously healthy men.

Over the next several months, additional similar cases were reported, and an apparent outbreak of other immunodeficiency-associated conditions was becoming noticeable. Kaposi's sarcoma—a very rare cancer in the United States, formerly observed only in elderly men and patients receiving immunosuppressive therapies—was now reported in young homosexual men in New York and California. As information was collected, it became apparent that twenty-six cases of Kaposi's sarcoma had been documented in thirty months: an alarming jump in the incidence of this cancer. A few of the patients with Kaposi's sarcoma also had *Pneumocystis* pneumonia and other opportunistic infections such as mucosal candidiasis (a fungal infection), disseminated cytomegalovirus infection (a latent herpesvirus), and chronic perianal herpes simplex virus-induced ulcers.

The common feature of the affected patients was that they all had evidence of T-lymphocyte dysfunction. Patients also exhibited a series of common characteristics—city of residence (predominantly Los Angeles, San Francisco, New York), age, race, and sexual orientation—and so an epidemiological study was carried out to compare patients with the disease to a similar group (matched for age, sex, race, sexual preference, and so forth) free of disease, searching for factors that predisposed one group to develop symptoms. The variable that was most clearly different between the patient group and the healthy group was the number and frequency of sexual contacts, which was much higher in the individuals with disease. By 1982, it was clear that a new disease entity had appeared, characterized by a severely impaired immune system and its related opportunistic infections and cancer. It was named acquired immune deficiency syndrome—AIDS.

Once a syndrome is defined, the search for a cause can proceed. The epidemiological studies indicated that some agent might be being transmitted through sexual relations, especially by sexually active men in the homosexual community. The sexual partners of several of the first AIDS patients in Los Angeles were contacted, and at least nine of them had had sexual relations with people who later—in some cases, five years later—developed *Pneumocystis* pneumonia, Kaposi's sarcoma, or both. In many cases, there was a long period of time between the sexual contact and disease symptoms recognizable as AIDS. The observation that several patients had a mild lymphadenopathy (an enlargement of the lymph nodes) and an abnormal immune response but few other symptoms began to suggest that the number of cases of AIDS in fairly asymptomatic individuals might be larger than was yet apparent.

By the end of 1982, it was clear that the outbreak of AIDS was not geographically limited, nor was it restricted to homosexual men. More than 800 cases were reported in 30 states; the populations afflicted included sexual partners of homosexual men, recipients of blood transfusions, hemophiliacs who received frequent injections of blood-derived products, intravenous drug abusers, and Haitian immigrants in the United States. In January 1983, the first well-documented cases of heterosexual transmission of AIDS were reported among partners of intravenous drug abusers. All these observations suggested a transmissible agent spread in genital secretions and blood, and several laboratories around the world began to receive samples of blood from AIDS patients to attempt an isolation of such an agent.

The Search for the Agent of AIDS Infection

About this time, Dr. Willy Rozenbaum of the Salpêtrière Hospital in Paris had a patient, a young French clothing designer who was homosexual, with lymphadenopathy. A sample of one of his enlarged lymph nodes was removed and examined histologically in an attempt at diagnosis; another sample was offered to Luc Montagnier, a virologist who worked at the Pasteur Institute, to see if an infectious agent might be isolated. The isolation of a virus from the lymph node of an early AIDS patient in the latent or asymptomatic period—should he prove to be one—was likely to be more significant than the isolation of a virus from an AIDS patient plagued with many opportunistic viral infections. Montagnier, working with colleagues Françoise Barré-Sinoussi and Jean-Claude Chermann, minced the node specimen and used the clear fluid from this tissue to inoculate cell cultures of lymphocytes from the umbilical cords of infants. He chose these lymphocytes as host cells because they were healthy, growing, young, and—since they derived from newborns—less apt to harbor viruses already. (In addition, a nearby obstetric hospital regularly supplied umbilical cords to his laboratory.)

To test for the presence of a retrovirus, a reverse transcriptase assay was used; the host cells do not have this enzyme, and it is fairly specific for retroviruses. After two weeks in culture and several negative results, samples with reverse transcriptase activity were detected. A retrovirus was present in the culture medium.

The previous experience with human retroviruses—HTLV-I had been isolated three years earlier, in 1980—had shown that cells infected with HTLV-I produced virus and were immortalized: they would grow in culture forever. But all attempts by the Pasteur group to grow these new infected lymphocytes in culture failed. The fact that no permanent cell line emerged from these studies meant that the amount of virus synthesized was limited; at the time, this seemed puzzling.

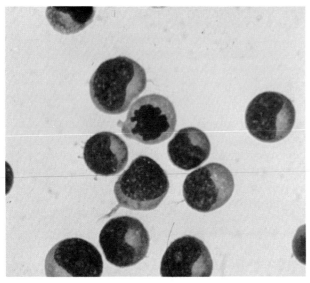

A

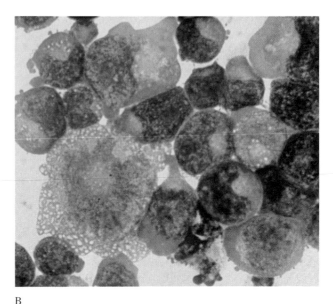

B

Light microscope views of T cells. A: Normal T-cell lymphocytes in culture are small, round cells with prominent nuclei. B: HIV-infected T cells often clump together, promoting the fusion of several cells into a large cell with multiple nuclei (a polykaryocyte). Such cells often have cytoplasmic vacuoles. With time, they die, releasing virus particles and reverse transcriptase that are detected in a test for the presence of HIV.

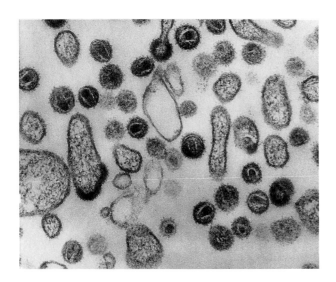

High-magnification EM of HIV particles.

The first thing to do was to determine if the retrovirus from this lymph-node biopsy was HTLV-I or HTLV-II. Antibodies against these viruses did not react with this retrovirus (HTLV-II was actually tested at a later date), so the Pasteur group named their virus lymphadenopathy-associated virus (LAV). Collaborating with David Klatzmann and Jean-Claude Gluckman of the Salpêtrière Hospital, the Pasteur group determined that LAV grew in CD4 T cells (helper T cells) but not in CD8 cells (killer T cells) and that it killed the host cell. Electron micrographs of LAV showed a retrovirus of the lentivirus group, quite distinct from HTLV-I or HTLV-II in structure.

In the meantime, by the end of 1983, samples of blood from AIDS patients were being tested for a retrovirus that could grow in T cells in continuous culture at the laboratory of Robert Gallo at NIH in

Bethesda. The NIH group succeeded in growing a virus from AIDS patients in a continuous culture of T cells, and they named their virus HTLV-III, for human T-lymphotropic virus-III. Growth in continuous culture permitted the large-scale production of the retrovirus and the preparation of reagents, such as antibodies directed against the virus, to detect it. The LAV isolate and HTLV-III were shown to be very similar to each other and related to other AIDS-virus isolates.

With the virus in hand, a test for antibodies that react with the gag and env proteins of this retrovirus could be carried out, and virtually all AIDS patients had such antibodies. These antibodies were not present in a random sampling of the population, but when high-risk groups (intravenous drug abusers, homosexual men with multiple sex partners, people who get frequent blood transfusions or blood products) were tested, a high percentage of these apparently healthy individuals had antibodies against this virus. As time went by, many of these individuals developed AIDS; that is, there is an asymptomatic latent period followed by clinical symptoms. These results confirmed the suspicion that the number of recorded AIDS cases constituted only a fraction of the people infected with the virus. Once these studies had clearly indicated that this retrovirus caused AIDS, it was renamed the human immunodeficiency virus (HIV).

Patterns of HIV Infection

The duration of the asymptomatic carrier state is unpredictable but can be seven to nine years or longer. The longest study comes from the San Francisco City Clinic cohort, in which 6700 homosexual and bisexual men have been interviewed, starting in 1978–1980 (this study began as a hepatitis-B program). After 88 months of HIV infections, 36 percent of the original asymptomatic carriers developed AIDS, 44 percent had clear signs of infection, and only 20 percent remained asymptomatic. The middle 44 percent are usually classified as AIDS-related complex (ARC) patients, to indicate the presence of two or more laboratory findings of immune dysfunction (these indicators vary in different groups—women, for example, or Africans). Once the diagnosis of AIDS is made, with all its symptoms, 50 percent of individuals will not survive for more than one year.

The HIV agent replicates in the CD4 T cells and kills them. Decreased numbers of CD4 cells in the blood are evident in AIDS patients, and a falling ratio of CD4 to CD8 cells (helper to killer T cells) is a common indicator of progressing disease. The CD4 T cells recognize foreign antigens in the body, help B cells make antibody, and help CD8 cells develop into killer cells (see Chapter 3). The loss of CD4 cells thus compromises both arms of the immune system—antibodies and killer T cells.

A normal individual will have between 800 and 1000 CD4 cells per cubic millimeter of blood. Asymptomatic carriers of HIV usually have 200 to 500 CD4 cells per cubic millimeter; at 200 CD4 cells per cubic millimeter, they are at risk for *Pneumocystis* pneumonia. Most *Pneumocystis* patients have between 100 and 200 CD4 cells per cubic millimeter. Some of the more common cancers associated with immunosuppression occur in patients with 50 CD4 cells per cubic millimeter, and reactivations of cytomegalovirus and CMV retinitis (eye infections) are seen in patients with 0 to 100 CD4 cells per cubic millimeter. Some AIDS patients have *no* detectable CD4 cells.

Worldwide, three patterns of HIV infection and transmission have been observed. Pattern I occurs in the United States, Mexico, Canada, many Western European countries, Australia, and parts of Central and South America. In these places, HIV probably began to spread extensively in the last half of the 1970s, with most cases in homosexual men and urban intravenous drug users. Transmission in the blood supply and in blood products was observed until 1985, but testing for the presence of HIV or antibodies has virtually eliminated this route of infection in these countries. Heterosexual transmission was rare but is now increasing, along with the transmission of HIV in utero. The spouses of high-risk individuals and prostitutes are the

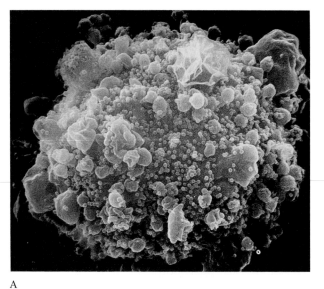

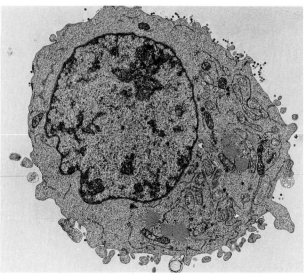

A

B

A: Scanning electron micrograph of the surface of an HIV-infected CD4 T-cell lymphocyte. The T cell with many surface projections has what is termed a ruffled membrane. The numerous small, circular projections are viruses budding from the cell. B: EM cross section of an HIV-infected CD4 T-cell lymphocyte. Virus particles (small, dark circles) budding from the cell surface and within vesicles in the cytoplasm are evident.

major sources of heterosexual transmission. Because of this transmission pattern, the male-to-female ratio of AIDS patients in these countries is about 10 or 15 to 1.

Pattern II is found in central, southern, and eastern Africa, the Caribbean, and particularly Haiti. These areas probably saw the most extensive spread of HIV in the late 1970s and early 1980s. Unlike pattern I, however, the predominant mode of transmission is via heterosexual contacts. Infected patients include equal numbers of males and females, and females in the 20-to-29-year age group have higher rates of HIV infection and AIDS than do males the same age. Male rates of HIV infection exceed female rates in the 30-to-39 and 40-to-49-year-old groups; infection rates are very low for both sexes below the age of 20 or above 50. HIV infection rates are high in urban prostitutes, and

there is a substantial amount of transmission to newborns. In many of these third-world countries, no tests of the blood supply are used, so blood-borne transmission remains a problem. The use of unsterile needles in the hospitals of some third-world countries also contributes to the spread of this virus.

With pattern III—typical of most Eastern European countries (except Romania, where newborns were routinely transfused with untested blood and infant HIV infection rates are very high), North Africa, the Middle East, Asia, and parts of the Pacific—HIV was probably introduced in the middle 1980s. In these regions, HIV is yet to be recognized as a major public-health problem. The virus was brought to these regions by contaminated blood products and sexual contacts with individuals from high-incidence areas of the world. The future course of AIDS in these countries will depend upon the ability of each government and population to respond to its distinctive health-care and educational needs.

The countries of central and eastern Africa, particularly the urban centers of the Congo, Rwanda, Tanzania, Uganda, Zaire, and Zambia, have been particularly devastated by AIDS. Although it has been difficult to obtain accurate epidemiological information in some of these countries, it has been estimated that 5 to 20 percent of the sexually active adults are already infected with HIV. Sixty to 88 percent of the prostitutes in Nairobi, Kenya, and Butare, Rwanda, appear to be HIV-positive. Almost 50 percent of the patients in urban hospitals are currently infected with HIV. By the middle of 1988, it was estimated that the cumulative number of cases in Africa was above 100,000; over the next five years, an additional 400,000 AIDS cases are expected. The existing health-care systems are simply not able to handle this projected rate of increase.

In 1986 a second human immunodeficiency virus, HIV-2, was isolated in West African patients. This virus is immunologically distinguishable from HIV-1 but shares some nucleotide sequences and antigens and has a similar gene organization. HIV-1 appears to have one gene not found in HIV-2, and HIV-2 has a different gene that is not seen in HIV-1. HIV-2 is certainly capable of causing AIDS in humans and has been isolated many times, but there is uncertainty about its relative virulence compared with HIV-1; HIV-2 may produce less severe disease and symptoms in humans.

HIV-2 is more closely related to a monkey retrovirus called simian immunodeficiency virus (SIV); when injected into Asian macaques in captivity, SIV causes a simian AIDS-like disease. SIV is not found in naturally occurring populations of macaques, however; rather, it is a virus of the African green monkey. In various natural populations of the green monkey, 30 to 70 percent of the animals are infected with this virus, which causes no known disease in these monkeys. SIV and HIV-2 are closely related immunologically (some antibod-

An African green monkey *(Cercopithecus aethiops).*

ies bind to both viruses), and they even share that extra gene found in HIV-2 but not in HIV-1. SIV and HIV-1 share about 50 percent of their nucleotide sequences, so these viruses are thought to be distant cousins. A retrovirus from the sooty mangaby, a monkey from West Africa, causes no known disease in its host but is 80 percent related by nucleotide sequence to HIV-2, suggesting a possible precursor for this human retrovirus. HIV-1 and HIV-2 have 45 percent correspondence (homology) of the nucleotide sequences over their entire RNA genomes.

All these observations have led some researchers to speculate that HIV-1 and HIV-2 arose in Africa from retroviruses of monkeys. Employing arguments about the expected rate of evolutionary change per generation for a nucleotide sequence in a virus, some scientists have gone on to suggest that the first human infections with HIV may have occurred more than twenty but less than one hundred years ago. Antibodies to HIV-1 have been detected in stored frozen blood samples taken from individuals in Zaire in 1959 and taken from patients in the United States in 1968, indicating the earliest dates we can be sure HIV was in circulation.

It has been postulated that rapid urbanization and social changes in Africa, with the disruption of traditional life styles, have brought HIV to the high-density cities of central Africa and promoted the spread of this virus; in this way, a virus that normally would arise and remain localized in a small, isolated group has broken out to infect a larger number of hosts. This is, of course, a very familiar story; social change and mobility have great consequences for the spread and impact of the parasites that prey upon us. Changes in the environment eliminate some and create great opportunities for others in the struggle for the fittest survivors.

The Molecular Biology of HIV

The genetic structures of HIV-1 and HIV-2 are presented in the figure below. The gag gene encodes a protein that is cleaved by the viral protease into four separate protein entities. The pol gene encodes the viral protease (pro), reverse transcriptase (RT), and integrase (IN), and the env gene produces a protein cleaved by a cellular protease into two glycoproteins, gp120 and gp41. Gp120 sticks out of the lipid envelope of the virus and makes contact with the receptor on the host-cell surface, permitting the initial attachment step. The cellular receptor is the CD4 protein, which is present on two types of blood cells: CD4 T cells (helper T cells) and monocytes. Monocytes give rise to macrophages, cells whose job it is to surround, envelope, and destroy invaders. These cells phagocytize foreign objects like bacteria or even dead cells and degrade them in their specialized vesicles (lysosomal vesicles and peroxisomes). The CD4 protein on the sur-

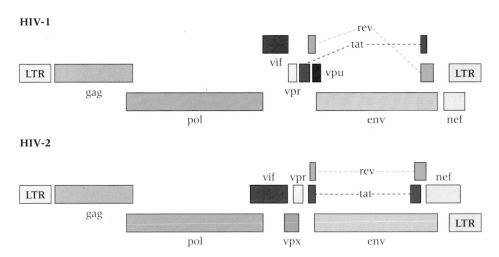

Genetic maps of HIV-1 (top) and HIV-2 (bottom). The LTR sequences at either end of the provirus surround the gag, pol, and env genes. Additional genes labeled tat, rev, vif, vpr, and nef are common to both viruses. When a gene is mapped to two noncontiguous regions of a chromosome, its mRNA is a spliced copy of each segment (dashed lines). Vpu is present in HIV-1 but not in HIV-2, while vpx is present in HIV-2 but not in HIV-1.

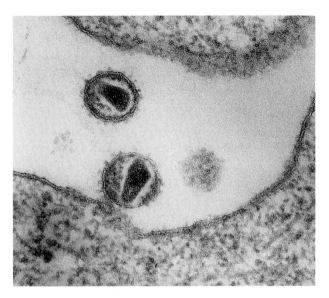

EM of HIV adsorption on a T-cell lymphocyte, mediated by the viral gp120 protein, which attaches to a CD4 protein on the cell surface.

face of the T cell and macrophage defines the host range of HIV-1 and HIV-2, which bind to this receptor protein. The viral gp41 protein is embedded in the viral lipid envelope and holds gp120 in the virion membrane.

Entry into the cell is not via endocytosis, as it appears to be for most retroviruses. Rather, gp41 promotes fusion of the viral lipid envelope and the host-cell plasma membrane into one continuous layer of membrane. This introduces the nucleoprotein viral core (RNA and gag proteins) into the cytoplasm. Reverse transcriptase copies the viral RNA into DNA, which then integrates into the cellular DNA. The provirus produces mRNAs for translation into the viral proteins, using the viral LTR as a promoter of transcription.

HIV-1 and HIV-2, which encode six additional viral genes along with the usual gag-pol-env structural and virion proteins, are the most complicated retroviruses studied to date. The tat gene appears to be similar to the HTLV-I tax gene, in that it positively regulates the level of viral mRNAs by acting

at a site in the LTR; when tat is made, more viral mRNAs are made and more virus is produced. A second viral gene, rev, is also essential for viral replication. Rev enhances the production or stability of gag-pol and env mRNAs. There is some evidence that rev helps to transport these mRNAs out of the nucleus and into the cytoplasm for translation into proteins. Rev appears to be functionally similar to the HTLV-I rex protein.

A third gene, nef, may act to negatively regulate (lower) the level of viral mRNAs and proteins. Some investigators, working with a nef-mutant virus, have found that HIV replicating in the absence of nef produces two to ten times more progeny. At first, it may seem strange that a virus would retain a gene whose function is to produce less virus. But reduced virulence is sometimes valuable: it keeps the host cell alive longer, so that—in the long run—more progeny overall are produced at a slower rate. Nef is not essential for virus replication in T cells in culture; it is useful in the host animal.

The fourth gene, vif (virion infectivity factor), is needed to spread the virus between cells. Viruses that do not make vif appear to produce normal amounts of viral mRNAs, proteins, and even virus particles. But when these viruses bud from the cell into the medium (in culture) or into the blood (in humans), they are not infectious. This virus can be transmitted from an infected to an uninfected cell via cell fusion, but the HIV does not spread through the environment. Vif appears to be important during the assembly of infectious viruses at the membrane, but the details of how it acts remain obscure. Two additional genes produced by these viruses are called vpr and vpu in HIV-1, vpr and vpx in HIV-2; the functions of their proteins are presently unclear, but they do not appear to be essential for virus replication in cell culture.

The infected T cell or macrophage begins to produce HIV, and virus buds from the plasma membrane. Inserted in the plasma membrane of infected cells are the gp120 and gp41 proteins, which protrude from the surface of these cells. When an infected CD4 T cell is close to an uninfected CD4 T cell or macrophage, the gp120 on the

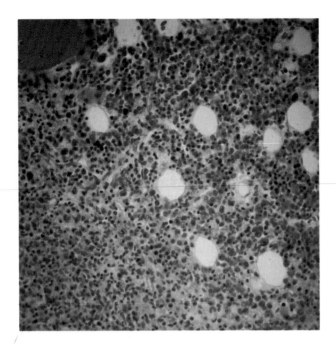

Light microscope view of a section through the bone marrow of an AIDS patient. In the bone marrow, where many different blood cells develop, it is common to find lymphocytes with large nuclei (blue). In this patient, some of the helper T cells are aggregated together because the gp120 of HIV cross-links cells via their CD4 receptors. This helps to spread the virus.

infected cell binds to the uninfected cell's CD4 protein, just as a virus particle might bind to this receptor; the gp41 protein then helps to promote fusion of the two cells. The plasma membranes merge, and one cell with two nuclei is formed. This process is repeated, forming a giant cell called a syncytium, with multiple nuclei. These syncytia are observed in the enlarged lymph nodes of people with lymphadenopathy. This is an unusual mode of transmission; a single infected cell fuses with many uninfected cells and efficiently delivers virus that has never left the original infected cell. In effect, normal cells take in an infected sibling, and the virus kills them all.

The infection of monocytes by HIV creates additional problems for the host. Monocytes are the precursors of several different cells in the body.

When they develop into macrophages, they function like filters: screening the blood for particles, dead cells, bacteria, or antigens and helping to fight infections. When monocytes enter specific tissues, they differentiate into anchored macrophages in the lung, Kupffer cells in the liver, Langerhans cells in skin, and microglial cells in the brain. It is thought that they then filter antigens and secrete chemical signals to attract immune cells (CD4 or CD8 T cells, or B cells) in these specific tissues. It is not totally clear whether HIV is carried to all these organs by monocytes or whether the virus can replicate in these monocyte-derived cell types (although microglial cells have been shown to be infected with HIV). What is clear, however, is that the virus does cross the blood-brain barrier that effectively prevents most substances from moving from the capillaries into brain tissue. An AIDS-associated dementia is a relatively common symptom late in the disease, and monocyte transmission of the virus to the brain has been suggested as a pathway to the central nervous system.

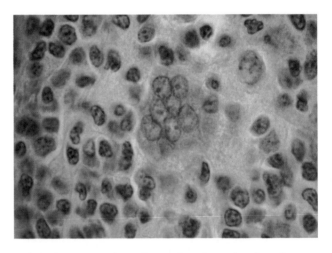

Light microscope view of a section through a lymph node of an AIDS patient. Lymph nodes are rich in both B and T cells. During HIV infections, gp120 cross-linkage of infected and uninfected CD4 lymphocytes results in cell fusion, creating multinucleate cells (polykaryocytes). Such a cell is shown in the center of the photograph.

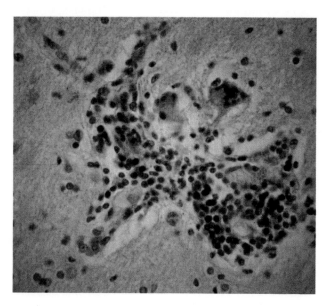

Chapter 7

HIV is brought to the brain by infected monocytes, which infiltrate through the blood vessels. This light microscope view of a section of a blood vessel in the brain of an AIDS patient shows lymphocyte infiltration and fused cells (polykaryocytes).

After the initial infection of humans by HIV, there is a long, variable asymptomatic period during which the virus replicates in CD4 T cells and macrophages but does not overwhelm its host. While there may well be an ever-increasing spread of the virus from cell to cell over time, this has no apparent impact on the function of the immune system for the overall health of the host. It has been suggested that environmental or perhaps genetic factors can accelerate virus production and the inevitable march to disease symptoms. For example, the addition of mitogens (agents that stimulate active division of T cells) to HIV-infected T cells in culture will result in dramatic increases in virus production and cytopathology. Foreign antigens may act similarly in people. Other viruses have also been shown to replicate in HIV-infected T cells, stimulating HIV production. The human herpesvirus type 6—and a new herpesvirus, type 7—infect CD4 T cells in this way and appear to help cell lysis

and HIV spread. Herpesvirus type 7 was originally isolated from an AIDS patient. Similarly, cytomegalovirus (CMV) stimulates HIV infections and has been found in HIV-infected brain tissues from an AIDS patient.

One of the unusual aspects of HIV-1 is the heterogeneity of the nucleotide sequences of independent isolates. When the nucleotide sequences of HIV agents are compared, some regions of the genome differ more than others; in general, the env gene and the nef gene are more variable than are the LTR, gag, pol, vif, vpr, tat, and rev genes. Over the total genome, HIV isolates can be up to 13 percent different in their nucleotide sequences—in the env gene, up to 30 percent different. Even the types of mutations in the env gene seem to differ from those seen in gag or pol. Deletions, inversions, and duplications of nucleotide sequences are all tolerated in the extracellular env gene and its product. The functional consequences of this rapid set of changes are not clear, but tissue and target-cell specificity, altered immunological reactivity, the clinical spectrum of disease development, and the host range of viruses could all be affected. It appears that isolates from Africa are often more divergent from each other than are American or European isolates. This has been said to support the notion that the virus has been in Africa longer than in the United States or Europe—using the additional time, so to speak, to evolve and change. Isolates of HIV-2 also show similar degrees of heterogeneity. SIV isolates within a given species—for example, the African green monkey—likewise show a great deal of diversity. The 10 to 13 percent changes in the HIV genome from different isolates can be contrasted with HTLV-I viruses from independent sources all over the world, which differ by only 1 to 4 percent.

Rapid changes in nucleotide sequences usually arise from errors made by the reverse transcriptase during the replication of RNA to DNA. These mutations permit the virus to try out many combinations of genes and proteins, since each infected cell has a provirus with slightly different nucleotide sequences. Natural selection then chooses from

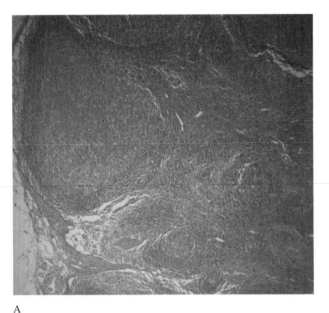

A

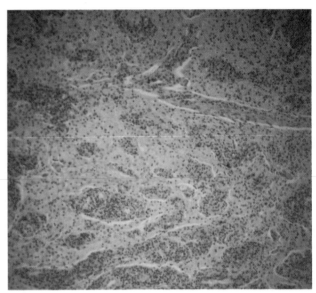

B

Light microscope views of sections through lymph nodes from HIV-infected patients. A: Early in HIV infection, a lymph node from an ARC patient is filled with B-cell and T-cell lymphocytes (bluish areas reflect large numbers of nuclei), densely collected in circular modules. HIV infection initially stimulates cell replication and consequent swelling (lymphadenopathy). This host-cell proliferation promotes virus replication and spread. B: A higher magnification view of a lymph node from an AIDS patient late in HIV infection shows marked depletion of lymphocytes (blue nuclei) as compared to an ARC patient (A). It is the continual loss of CD4 helper T cells that causes failure of the immune system and death.

among all the mutants for the fittest virus. A high mutation rate can result from the built-in error frequency of a virus's reverse transcriptase, as well as from the plasticity of its proteins. We can speculate on just what advantages or disadvantages this imparts; the real answers will come as we watch this virus and disease evolve and respond to the drugs and vaccines we place in its path.

Strategies for Combating HIV and AIDS

There have been a number of different approaches to designing drugs, vaccines, and treatments to address the problems of HIV infection and AIDS. The only drug presently licensed for use in the treatment of AIDS that is designed to kill the virus is 3'-azido-3'-deoxythymidine, or AZT (also called zidovudine or retrovir). AZT is a nucleotide analog of thymidine (that is, it has a similar structure). The analog uses the cellular thymidine kinase to add a phosphate group to AZT, and AZT triphosphate is incorporated by the reverse transcriptase into viral DNA. When present at the end of a growing chain of DNA, it stops further DNA synthesis. AZT triphosphate inhibits the synthesis of DNA by reverse transcriptase about a hundred times better than it inhibits the synthesis of DNA by the host-cell DNA polymerase in the cell nucleus. Moreover, AZT inhibits HIV replication at concentrations about a thousandfold less than it takes to inhibit the replication of the host-cell lymphocytes. This differential toxicity led to the clinical testing of AZT in AIDS patients.

Between February 1986 and June 1986, 282 patients were enrolled in a study to determine whether AZT taken orally would improve the outcome of AIDS. In such studies—termed double-blind, placebo-controlled clinical trials—patients are randomly chosen and coded to be in one of two groups, one given the test drug and the second given a placebo; no one in the clinical setting, doctor or patient, knows who is in which group. But if the drug is really working well, it is desirable to get it to the placebo group quickly. For that reason, an independent data-safety monitoring board breaks the code and terminates the trial early if the efficacy of the drug is proven rapidly. By September 1986, this study was terminated by the board. Nineteen patients (twelve with AIDS and seven with ARC) in

the placebo group had died, compared to only one patient who had received AZT.

AZT extends life, reduces opportunistic infections, and improves the immune response in many patients. It is not without side effects, however, and these have caused some patients to withdraw voluntarily from AZT treatment. In addition, HIV strains that are resistant to AZT can arise via mutations in the reverse transcriptase gene and the protein it encodes. Several other nucleotide analogs—such as dideoxycytidine, which acts much like AZT—are being tested now for their ability to kill the virus and reverse the disease.

A second approach to drug therapy is to inhibit the viral protease required for cleavage of the gag and env proteins. The gene for the HIV protease has been placed in bacteria, and with the large amounts of protein made that way, it has been possible to work out rapidly the three-dimensional structure of this viral protein. Several pharmaceutical companies have announced that they are working to design drug analogs to fit into the protease and inhibit

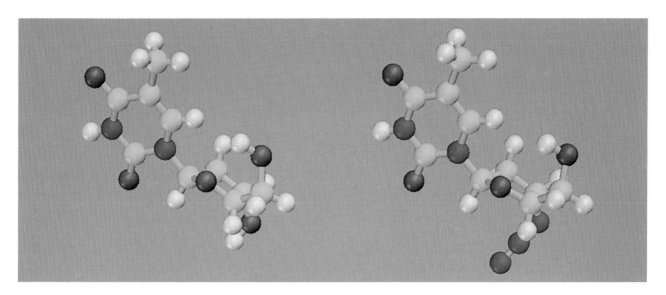

Computer-graphic representation of the structures of thymidine (left) and AZT (right). AZT acts to inhibit DNA synthesis by HIV, inducing the reverse transcriptase enzyme to use it instead of thymidine.

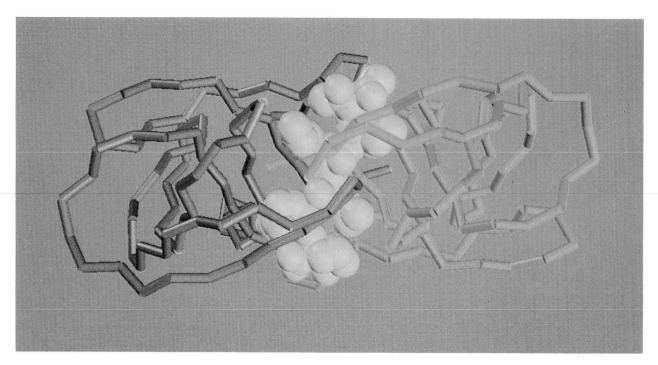

Computer-graphic representation of the structure of the HIV protease, based on X-ray diffraction patterns from crystals of the polypeptide chain. The two identical proteins of this dimer (here shown in orange and green) are symmetrical about an axis where the gag and pol cleavage reaction occurs (the active site). Inhibitors of this process (here, blue spheres in the active site) have been designed.

it from cleaving the gag and env proteins in an HIV-infected cell.

It would be very useful to have two drugs that have different viral proteins as targets. Say the virus is able to mutate at a frequency of once in a million viruses so as to create an AZT-resistant reverse transcriptase; and further hypothesize viral mutation for a protease-inhibitor resistance at one in a million viruses produced. The chances, then, that one virus would contain both mutations for drug resistance is one in a million million (multiply the probabilities of independent events)—one in 10^{12}. That is a lot of viruses. If two drugs are given simultaneously to a patient, the chance of a single virus arising that is resistant to both drugs is far smaller than the chance of a virus becoming resistant to either drug alone. This is why we do not give drugs sequentially: resistance arises to each drug separately at one per million, while resistance to two drugs administered simultaneously occurs once per 10^{12}. There are other, less obvious targets for drug therapy (tat, rev, integrase) but much less is known about how these proteins function and how to inhibit them.

A different approach has been to synthesize a large amount of the CD4 protein receptor. This has been done by placing the human CD4 gene, or a part of it, into bacteria and stimulating the bacteria to synthesize large amounts of a soluble, non-membrane form of the portion of CD4 that binds to the viral gp120 protein. Placed in the blood of AIDS patients, this soluble CD4 binds to gp120 in virus particles and blocks or competes with the viruses' attachment to cells; in this way, viral infection is reduced. Unfortunately, the amount of CD4 needed to block all viral attachment to cells is great, and the body degrades it too rapidly. A variation of this approach has been to link the soluble CD4 protein to a toxin that kills cells. HIV-infected cells have gp120 on their surfaces, so the soluble CD4, binding to gp120, would bring the toxin specifically to an infected cell. Experiments like this one are being attempted now.

Although a few drugs have been effective in fighting viruses, the way we have controlled and conquered viruses in the past is through safe and effective vaccines to prevent both disease and viral infections, by maintaining a high level of immunity in populations. In the past, the best vaccines have been attenuated live vaccines because they replicate and provide a strong immune response, giving us the closest thing possible to lifelong immunity. We are simply not sure how to attenuate HIV. There are no animal models in which to test an attenuated virus because no animal shows the symptoms of AIDS after HIV infection. What mutations would permit the virus to replicate, but not to kill the host cell?

Even if we produce such an attenuated virus, we still must worry about the high HIV mutation rate, which could give rise to revertant viruses that again kill the host cell. Recall that our human DNA harbors retrovirus nucleotide sequences from our past as a species. Could an attenuated HIV recombine with these retroviruslike sequences in our chromosomes to form a new virus? Introducing a live, even attenuated, retrovirus into the human population could create many unintended problems. If millions of people were vaccinated with such an agent, rare events could result in the creation of new viral agents. For these reasons, a live attenuated virus is probably not our first choice for a vaccine.

That leaves a killed-virus vaccine or a subunit vaccine composed of protein only. Some virologists would prefer to stay away from a killed intact virus, which still contains reverse transcriptase and HIV RNA; introducing this into people creates a chance for a DNA provirus to be made with unknown, possibly adverse, consequences to the host. A purified gp120 protein would eliminate this objection. Work on such subunit vaccines is in progress, but this approach is not without its difficulties. The portion of the gp120 protein that binds to the CD4 receptor, which has been identified and studied in some detail, appears to form a pocket. Therefore, there is concern that this pocket-shaped site—where antibody binding to gp120 would block its ability for cell attachment—might be poorly immunogenic (weak at eliciting an immune response).

A second problem is that gp120 seems to tolerate a very high mutation rate, with up to 30 percent variability between gp120 proteins from different isolates of HIV. While these wide differences do not occur in the gp120 pocket that binds to CD4, it is unclear how the mutations seen in gp120 might affect antibody binding in or around the pocket. The high mutation rate could produce some viruses that escape our antibodies; these would be rapidly selected for in an immunized population, and we would be back at the starting line.

Third, there may be a problem with stimulating high levels of anti-gp120 antibodies in human beings. When an HIV agent infects an immunized individual, the antibodies will bind to gp120 on the virus particle, preventing it from attaching to CD4 receptors. Now the body must get rid of the antibody-virus complex, which it does by secreting soluble complexes in the urine and also by using macrophages to collect the complexes and degrade the antigen; macrophages have a special receptor for this purpose. Under certain laboratory conditions, at least, low levels of antibody against HIV actually enhanced entry of HIV-1 into monocytes or macrophages. We know that these cells will replicate HIV and produce more virus, if the HIV-1 agent escapes the lysosomes of the macrophages. In this case, antibody to gp120 could well provide an alternative route to a permissive cell.

Finally, transmission of HIV by cell fusion, where one infected cell can recruit uninfected cells, might not be blocked by antibodies or effectively

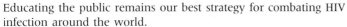

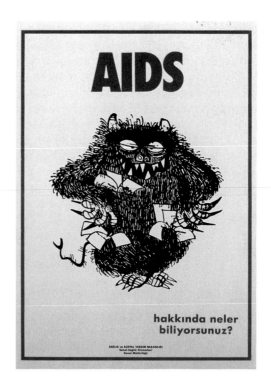

Educating the public remains our best strategy for combating HIV infection around the world.

reduced by neutralizing antibodies—during cell fusion, HIV does not have to enter the bloodstream, where antibodies are located.

Despite these theoretical objections to a subunit gp120 vaccine, work in this area is progressing and should continue. It simply is not possible to predict whether these difficulties will in fact arise. In addition, a vaccine that could stimulate the CD8 (killer) T cells, which might well recognize different viral proteins (gag, for example), will be an important alternative approach to stimulating humoral, or antibody, immunity.

Another attractive approach to a subunit vaccine is to insert the HIV gene for gp120 or other viral proteins into the chromosome of a *different* live attenuated virus, which could then be used to immunize a host; by virtue of its own replicative abilities, it could provide a strong and long-lasting immune response against gp120. Among the best candidates for this so-called recombinant virus vac-

cine (recombining the gene for gp120 into the carrier-virus chromosome) is vaccinia, the vaccine strain for smallpox. A variety of experiments have shown that this viral chromosome can accept and express foreign genes, and animals immunized with this virus develop antibodies to the foreign protein. Indeed, vaccinia viruses expressing several HIV proteins have been made and are being tested in animals. Vaccinia has a long, safe history in humans, but its use in the population has been discontinued because the smallpox virus has been eradicated (see Chapter 3). Perhaps HIV will be the stimulus to bring back vaccinia as an immunizing vector in humans. In any case, developing and testing the efficacy of a safe vaccine will take years. Its introduction lies in the future.

What has been effective to date is preventing the spread of this virus from individual to individual by understanding and addressing the primary modes of transmission. The development of screening tests for HIV antibodies has improved safety in the blood supply in the West. It is critical to introduce these tests in Africa. Recommendations for behavior modifications among high-risk groups, such as homosexual men with multiple partners and intravenous drug users, are essential in stopping the spread of this virus. Education of members of high-risk groups is beginning to pay off. The rate of increase of HIV cases in homosexual men declined in 1990 because of excellent educational programs. While the gay community has responded to this crisis in a comprehensive and politically active fashion, no comparable organization exists for intravenous drug abusers. The consequences of pregnancy for women whose serum contains antibodies to HIV need to be understood and alternatives offered, since a variable percentage of such women will pass the virus to their offspring during pregnancy. An enormous effort is required worldwide to continue education of the public and to inform schoolchildren about AIDS and HIV before they become sexually active.

The Influenza A Virus

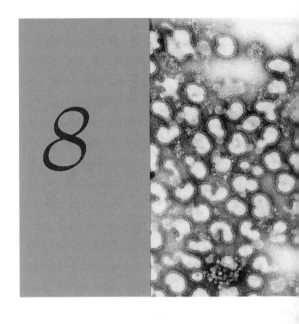

In the space of one hundred and seventy-six years the lower Mississippi has shortened itself two hundred and forty-two miles. That is an average of a trifle over one mile and a third per year. Therefore, any calm person, who is not blind or idiotic, can see that in the Old Oölitic Silurian period, just a million years ago next November, the lower Mississippi River was upwards of one million three hundred thousand miles long, and stuck out over the Gulf of Mexico like a fishing rod. And by the same token, any person can see that seven hundred and forty-two years from now the lower Mississippi will be only a mile and three-quarters long, and Cairo, Illinois, and New Orleans will have joined their streets together and will be plodding comfortably along under a single mayor and a mutual board of aldermen. There is something fascinating about science. One gets such wholesome return of conjecture out of such a trifling investment of fact.

Samuel L. Clemens (Mark Twain), 1875

Left: On the evening of October 27, 1918, the artist Egon Schiele sketched this last portrait of his wife, Edith; she died the next morning, as the century's greatest pandemic (worldwide epidemic) of influenza A virus swept through Vienna.

Above: EM of influenza A virus particles, each surrounded by a lipid envelope with protruding glycoprotein spikes. Rapid evolution in these surface antigens is responsible for the emergence of fatal pandemic strains.

All viruses require a portal of entry into the host, a specific tissue where they can set up a replicative factory. A virus from the external environment, approaching its potential host, is presented with a set of surfaces that face the outside world. In humans, these include the skin, mouth, digestive tract, respiratory tract, and urogenital organs. We have now reviewed viruses that attack each of these surfaces. Herpes simplex virus type 1 and the papilloma viruses enter through our skin. EBV first replicates in the nasopharynx, and poliovirus initially infects the epithelial lining of the intestine. HTLV-I, HIV, and HSV-2 all are transmitted to the surfaces of the urogenital tract during sexual intercourse. The tonsils, adenoids, and cells of the respiratory tract are the targets of the adenoviruses.

Within the respiratory tract, the extensive surface areas of the lung have attracted a particularly large share of parasites, which set up both chronic and acute infections resulting in symptoms that range from mild respiratory distress to pneumonia. The term *influenza* has been used for centuries to describe respiratory infections caused by bacteria, viruses, fungi, protozoans, or other pathogens within the lungs. As infectious agents were isolated and identified with symptoms, they were often named for the disease they caused; in this way, several bacteria and a few viruses came to be called ''influenza.'' At present, there are three viruses with this name—influenza A, B, and C—all of which cause respiratory disease. We will focus on the influenza A virus because it causes the most clinically significant disease and because its unique properties permit it to cause epidemics and to infect the same host more than once during a lifetime.

Epidemics of viral diseases most often result from some unusual ecological perturbation—for example, the introduction of a large number of susceptible people into an area where a disease is endemic, as may happen during a war or its aftermath. Alternatively, rampant disease has followed the introduction of a new virus into a population that had no previous exposure, as when smallpox devastated the Amerindians and measles, the Inuit. Yet a third pattern can be noted with poliovirus, which was endemic to many regions but became epidemic when improved sanitary conditions and public health measures increased the numbers of susceptible hosts.

Viral disease is usually endemic, with periodic fluctuations of intensity in response to ecological events and population dynamics. Influenza A, however, has two epidemic patterns, which are fairly atypical. Both patterns occur in the fall and winter. The first type sweeps through defined geographical locations yearly or every few years—appearing suddenly, persisting for a few weeks, and disappearing rather abruptly. By contrast, at irregu-

War and pandemic, 1918.

lar intervals years apart, influenza A traverses the world to create a pandemic—a worldwide global epidemic. Commonly starting in Asia, moving through Russia into Europe, and then spreading to the Americas, these pandemics can result in high mortality among elderly people. Such a pandemic was first described in 412 B.C. by Hippocrates as it passed through Greece (presumably on its way to western Europe).

The previous chapters have made it clear that our response to virus infections is to mobilize the immune system, inactivate the virus particles with antibodies, and use killer T cells to destroy the infected cells. This leads to a recovery from the disease and to lifelong immunity—which prevents reinfection and severely limits the number of hosts available to a virus. If an agent can successfully infect a host only once in a lifetime, that virus needs good mobility to get to as many different people as possible. In response to this limitation, viruses have evolved strategies to overcome or bypass the immune system. We have seen how HSV-1 and HSV-2 hide in neurons, remaining in the latent state within cells where antibodies cannot reach them. Other viruses persist in the body at low levels (adenoviruses) or even at the high levels characteristic of chronic infection with hepatitis B virus, as we shall see in the next chapter. These viruses come to equilibrium with the immune system of the host: neither wins decisively. The immune response continues to function, which keeps the host healthy overall, but the particular virus retains its replicative base by synthesizing proteins that compromise the ability of T cells to recognize infected cells as foreign. A final group of viruses (HIV) simply destroy the immune system of the host.

The influenza A virus has adopted none of these approaches. Rather, it has developed the ability to evade recognition by the immune system's memory. During replication, it sometimes alters the viral structural proteins rapidly but slightly through mutation—just enough so that the less efficient immune systems of some hosts no longer recognize the progeny virus as the same. If the immune system has no memory of a past infection, each infection is a new exposure and the virus can gain access to the host multiple times. This process, called antigenic drift, is responsible for local epidemics.

Quite a different mechanism comes into play when a pandemic influenza A virus sweeps around the world. In this case, the virus replaces one of its genes with a new one, producing a protein that the immune system has never seen before. This radical change, termed antigenic shift, produces a virus that is still influenza A but with one or more gene replacements. No longer recognized as an agent identified in the past by the immune system of any host, this virus can infect the same population one more time. This virus seems, in effect, to have an interchangeable set of functional genes lying mysteriously in storage somewhere—perhaps in China. How influenza A evolves so rapidly is the story of this chapter.

The Structure of Influenza A Virus

Influenza A virus is an RNA virus. Each virion contains eight RNA molecules—eight chromosomes—termed segments 1 through 8, which vary in length from 890 to 2341 nucleotides. Segments 1 through 6 are each one gene long and encode the information for a single protein; segments 7 and 8 comprise two genes, encoding separate proteins. Thus, eight RNA molecules encode ten proteins, eight of which are found in the virus particle and two of which

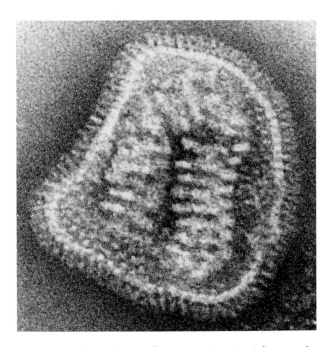

EM section through an influenza A virus particle, revealing the helical ribonucleoprotein core surrounded by a viral-protein matrix and the lipid membrane incorporating HA and NA subunits that protrude like spikes.

	RNA length		Protein length	Approximate number of protein molecules
Segment	(nucleotides)	Protein	(amino acids)	per virus particle
1	2341	PB2	759	30–60
2	2341	PB1	757	30–60
3	2233	PA	716	30–60
4	1778	HA	566	500
5	1565	NP	498	1000
6	1413	NA	454	100
7	1027	M1	252	3000
		M2	97	
8	890	NS1	230	
		NS2	121	

Influenza A Virus RNA and Proteins

(NS1 and NS2) are synthesized in an infected cell but never enter the fully assembled virion.

In the core of the virus, each of the eight RNA molecules forms a complex with molecules of NP (the nucleoprotein). Each virus particle has about 1000 NP molecules, so one NP covers about twenty nucleotides, packaging this RNA into a helical core structure that can be visualized in some electron micrographs. The NP subunits seem to protect the RNA from the environment and to facilitate its duplication.

Surrounding the ribonucleoprotein core is a structural layer of matrix protein, M1; this protein is present in large amounts—about 3000 molecules per virion particle. Resting on the matrix is a lipid coat that is derived from the plasma membrane of the cell when influenza A virus, much like the retroviruses, buds from the cell surface. Inserted into this membrane are viral proteins called the hemagglutinin (HA) protein and the neuraminidase (NA) enzyme, plus a few molecules of the M2 protein, the function of which remains unclear.

The HA protein got its name because it can bind to certain sugars on the surface of red blood cells, coating the cell surfaces. This process, called hemagglutination, causes the red blood cells to clump together when the viruses form a bridge between them. Red blood cells attract the virus, which attaches to them, but they lack the nuclei necessary for viral replication. The nucleated epithelial cells of the lung, however, have similar surface receptors, and influenza A virus particles adsorb to them by means of the HA protein.

The structure of the HA protein has been studied in some detail. Each protein folds into the shape of a lollipop, with a stem and a ball. The fibrous stem is anchored into the membrane of the virus particle. Each HA molecule associates with two others, forming a trimer of stems and globular spheres held together by chemical forces in structural threefold symmetry. In a protein pocket near the top of each sphere is the recognition site where HA binds to its sugar, which is called sialic acid (neuraminic acid).

Neuraminidase, which also protrudes from the viral membrane, is an enzyme composed of four molecules that form a box-shaped tetramer with fourfold symmetry. A slender stalk attaches this tetramer to the particle membrane. NA has the ability to cleave off a sialic (neuraminic) acid molecule from the surface of a cell or particle. While the functions of NA are not totally clear, this cleavage can remove the cell receptor bound to an HA molecule, promoting release of the virus from a cell; conversely, it can cleave the receptor from an already infected cell, and so prevent redundant virus attachment. Such a mode of action also reduces the aggregation of viruses on mucus particles, which are coated with sialic acid and are common in the respiratory tract. The NA subunit also removes sialic acid from the HA protein itself, preventing self-aggregation of virus particles. During the maturation of the virus, HA is cleaved by a protease into two subunits (much like the env proteins of retro-

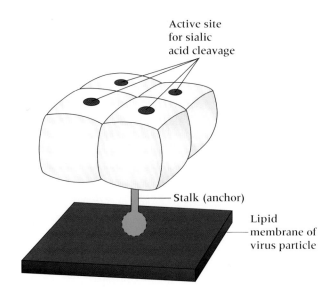

Schematic representation of the neuraminidase protein tetramer on the surface of the influenza A virus: the NA spike.

viruses), and it is possible that the NA's removal of sialic acid from HA assists this process.

Each virus particle contains about 500 molecules of HA (170 trimers) and 100 molecules of NA (25 tetramers) on its lipid surface. HA covers the viral surface uniformly, while the NA tetramers tend to cluster together in patches. Influenza A viruses are commonly irregular spheres about 120 nm in diameter. Filamentous virus particles, composed of long tubes of membrane surrounding helical cores of ribonucleoproteins, are also seen in the electron microscope.

The Replication of Influenza A Virus

An influenza A virus particle adsorbed by its HA to sialic acid on the surface of a cell enters the cytoplasm by endocytosis. A cell coated with virus particles will take in half of them every ten minutes through this normal process of vesicle formation,

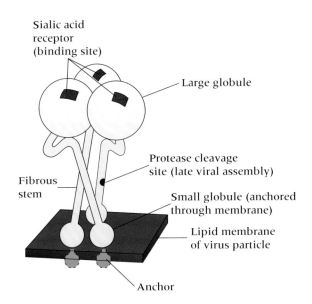

Schematic representation of the protein trimer that forms the hemagglutinin antigen on the surface of the influenza A virus: the HA spike.

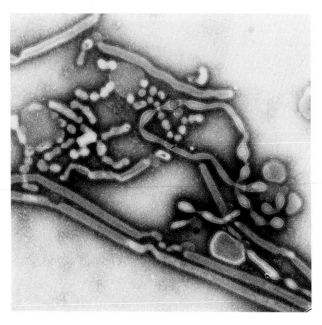

EM of filaments and spheres, the most common forms of influenza A virus. Depending upon its environment, the virus particle can take many shapes. Filamentous particles may have many copies of the viral chromosomes, ensuring high infectivity and reproductive efficiency.

so infection occurs rapidly and efficiently after adsorption. The vacuoles containing virus particles are delivered to endosomes, acidic vesicles in the cell. In this low-pH environment, the HA trimer changes its shape, probably exposing a portion of the protein in the stem of the "lollipop," so that it can touch the vesicle membrane. This region of the stem is called the fusion peptide, since it is thought to fuse the viral membrane with that of the lysosomal vesicle. The net effect is to release the ribonucleoprotein core of the virus into the cytoplasm and ultimately into the nucleus, where the next stage in this replicative drama takes place.

In the nucleus of an infected respiratory cell, thousands of cellular genes of the DNA chromosome are being transcribed into mRNAs, which are processed and shipped out to synthesize proteins on the cytoplasmic ribosomes. The influenza A ribonucleoprotein core "steals" short pieces of these mRNAs from many different genes, cutting them off 10 to 13 nucleotides from the start site of RNA synthesis. These pieces of cellular RNA are used as primers for the synthesis of influenza mRNAs, which are now manufactured using the eight RNA chromosomes as templates.

Because the RNA genome of the influenza A virus has negative polarity, it must be copied into positive-strand RNA that can be translated on the ribosomes of the infected cell to synthesize viral proteins. Transcribing the RNA is the function of three proteins—PB2, PB1, and PA—that are closely associated with the NP-RNA complexes. These proteins form an RNA transcriptase. Beginning from the starting sequence stolen from the cellular mRNA, the viral polymerase complex appears to extend the mRNA by synthesizing a hybrid molecule—part host (primer), part viral (extension). Because PB2, PB1, and PA are enzymes and can act repeatedly to catalyze these reactions without being used up, only a few molecules per virus particle need be present to produce lots of RNA.

The PB2 protein involved in stealing the required primer sequences from the cellular mRNA simultaneously inhibits the ability of the cell to make its own proteins. The virus maximizes production of the ten viral proteins by cutting off the mature cellular mRNAs, preventing them from ever getting to the ribosomes. Instead, the finished influenza mRNAs move to the ribosomes, where they are translated.

Proteins like HA, NA, and M2 contain the ZIP codes to pass into the endoplasmic reticulum, from which they are inserted into the plasma membrane. The NS1 and NS2 proteins both return to the nucleus, but their functions are still unclear. The viral structural proteins—NP, M1, PB1, PB2, and PA— are all returned to the nucleus after synthesis.

At this time, after viral-protein production, the positive-strand copies of the viral chromosomes previously employed as mRNAs switch function, becoming templates for the synthesis of the eight negative strands of the RNA genome—again by the action of the PB2-PB1-PA RNA transcriptase. The

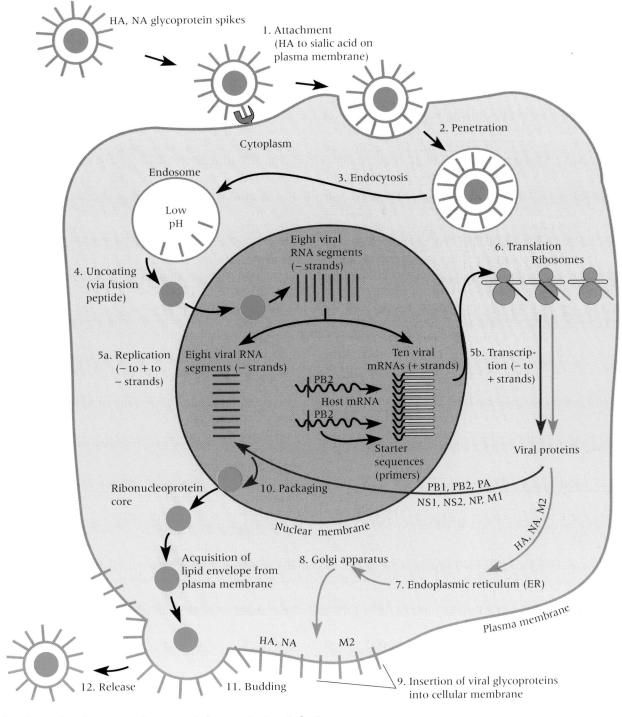

HA, NA glycoprotein spikes

1. Attachment
(HA to sialic acid on
plasma membrane)

Cytoplasm

2. Penetration

Endosome

3. Endocytosis

Low
pH

Eight viral
RNA segments
(− strands)

6. Translation
Ribosomes

4. Uncoating
(via fusion
peptide)

5a. Replication
(− to + to
− strands)

Eight viral RNA
segments (− strands)

Ten viral
mRNAs (+ strands)

5b. Transcrip-
tion (− to
+ strands)

PB2

Host mRNA

PB2

Starter
sequences
(primers)

Viral proteins

Ribonucleoprotein
core

10. Packaging

PB1, PB2, PA
NS1, NS2, NP, M1

HA, NA, M2

Nuclear membrane

Acquisition of
lipid envelope from
plasma membrane

8. Golgi apparatus

7. Endoplasmic reticulum (ER)

Plasma membrane

HA, NA   M2

12. Release   11. Budding

9. Insertion of viral glycoproteins
into cellular membrane

Flowchart of major events during an influenza A virus infection.

eight viral chromosomes, duplicated into thousands of copies, are packaged into ribonuclear particles by the NP protein. A small number of PB2, PB1, and PA molecules are added, and the M1 matrix protein covers this ribonucleoprotein core.

The core finally migrates from the nucleus to the plasma membrane, where NA and HA subunits have already been inserted, and buds out of the cell, taking its lipid envelope with the embedded NA and HA proteins. (A critical late event in triggering infectivity is the cleavage by protease of the HA protein, a step roughly parallel to one we observed in the retroviruses and essential in viral assembly and release.) The shedding of progeny virus goes on for hours until the infected cell dies.

From the standpoint of the virus, packaging of the influenza A genome is a crucial process. In a cell nucleus that contains tens of thousands of newly copied viral chromosomes, it is not at all clear how each core particle collects and packages the eight different subunits. Obviously, it would not do for the virus to package eight identical chromosomes (NA, for example); the virus needs all ten different genes to duplicate itself in the next round of replication. As will become clear as this chapter progresses, proper chromosomal segregation holds the key to the future of different influenza strains and their ability to cause pandemics.

The Epidemiology of Influenza A

In the late 1920s, Richard Shope isolated the first influenza A virus, from pigs. He was able to show that a solution of filtered mucus from infected animals reproduced the disease in other pigs. Because the filters used were not large enough to pass bacteria, the influenza A agent had to be a virus. In 1933, influenza A virus was first isolated from humans by a team at the National Institute for Medical Research in London. Since those original observations, it has become clear that the same influenza A viruses, or similar ones, can be found in diverse animals including humans, swine, horses, a wide variety of birds, and even seals. As a large number of influenza A viruses were isolated from several different species, it became critical to identify them and classify each isolate. Scientists were then able to ask: Is this virus the same in two different species, or even in two different individuals, or is it a different virus?

As we have seen, antibodies made by a host against a virus are very specific. We are only protected by antibodies against a virus we have encountered in the past. For that reason, antibodies from humans or animals exposed to influenza A virus were used to test new isolates and establish whether each was the same as or different from other influenza A subtypes. While antibodies are made against many of the influenza A viral proteins, the two most informative are the HA and NA antigens, the proteins found on the surfaces both of virus particles and of the plasma membranes of infected cells. Furthermore, antibodies directed against the HA protein can prevent the virus from infecting cells. These so-called neutralizing antibodies are thought to prevent the HA molecule from attaching itself to the cell's sialic acid receptor.

When animals or humans produce antibodies that recognize and bind to a foreign antigen, the host antibodies recognize specific sites on the antigen surface, called epitopes. The antibody recognizes the chemical composition (electrical charges) and shape of the epitope, both of which are determined by the protein's amino-acid sequences. Antibodies that neutralize the influenza A virus do not do so literally by binding to the HA receptor site, thereby competing against the combination of viral HA and cellular sialic acid, since this site is protected by a pocket too small to contain an antibody molecule. Rather, there are five epitopes on or near the sphere of the HA molecule that are now known to bind antibodies, which then neutralize the infectivity of this virus. In the adjacent figure, they are termed sites A, B, C, D, and E.

Epitopes A, B, and D surround the HA pocket, so one might imagine that an antibody binding to one of these three locations could be large enough to block adsorption to sialic acid. Sites C and E ap-

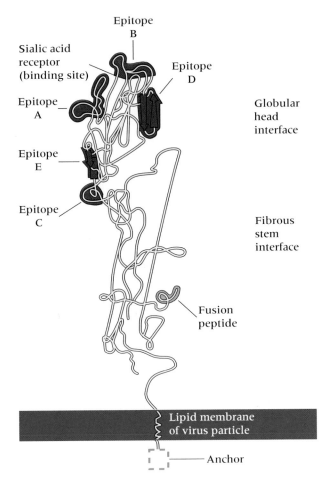

Epitope
B

Sialic acid
receptor
(binding site)

Epitope
D

Epitope
A

Globular
head
interface

Epitope
E

Epitope
C

Fibrous
stem
interface

Fusion
peptide

Lipid membrane
of virus particle

Anchor

Approximate structure of the protein chain of a single subunit of the influenza A HA trimer. Epitopes (sites) A, B, C, D, and E are the positions where antibody molecules bind to the HA molecule. The three-dimensional structure was determined by X-ray diffraction of crystalline protein.

pear to be too far from the pocket to act in this fashion, so some other mechanism must be postulated. Recall that as the virus enters the cell in an endocytic vesicle, the acidic environment (pH 5.0 or so) induces a dramatic change in the shape of the HA protein. Antibodies that bind to sites B and D at normal pH fail to do so at pH 5.0—which means that sites B and D move, under acidic conditions, from the surface to the inside of the molecule, where antibodies can no longer reach them. It is certainly possible that an antibody already bound to site B or D may prevent such a rearrangement. In the absence of the pH-induced conformational change, the fusion peptide in the stem of the HA molecule may not be released to associate with the plasma membrane. As we have repeatedly observed, events in the life cycle of a virus must proceed in a chronological order in response to environmental changes, and interference in that order can disrupt the cycle. (Similar epitopes found on the NA protein of the influenza A virus do not lead to its neutralization, but are useful nonetheless in classifying subtypes.)

As thousands of isolates of influenza A viruses were obtained, a number of distinct subtypes came to be recognized. Two kinds of antibody-binding patterns were seen. Sometimes HA molecules from two different viruses differed at one or two epitopes (site A or C, for example), but were similar at other sites (B, D, and E). At other times, most or all of the five sites were different in two isolates. In such a case, the agent was still influenza A virus: it had an HA that bound sialic acid, an NA, and all the other viral proteins. The only difference was that its HA failed to bind to antibodies produced against the HA protein from a different subtype. Each time an antibody was found that was specific for a different HA subtype, it was given a new number—H1, H2, H3, and so forth in series to H14, the present number of immunologically and genetically distinct HA types. Each time an antibody specific for an NA subtype was produced, it too was given a number from a similar series: N1 to N9, the present number of distinct subtypes. Different virus isolates from humans, birds, seals, and so on could then be characterized as, for example, H1N1 or H2N6.

Because so many different viruses were being identified as influenza A subtypes, a uniform and informative nomenclature was invented to describe each agent. Conventionally, the name of an isolate includes (1) the letter A for influenza A; (2) the host of origin, unless the virus is isolated from

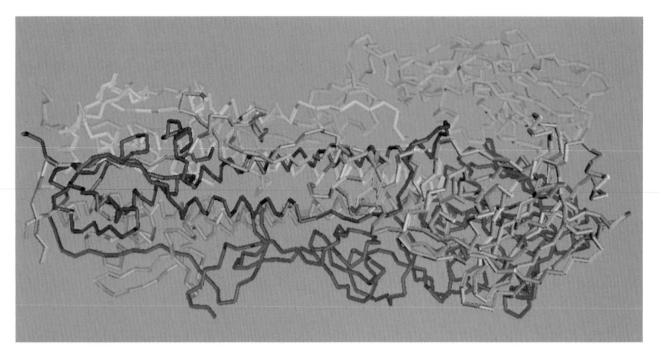

Computer-graphic representation of the HA trimer. Blue, yellow, and green each trace the polypeptide chain of one of the three HA monomers. Anchor is at left; globules are at right.

humans, in which case no indication is given; (3) the geographical location of the isolate; (4) a strain number followed by the year of isolation; and (5) the HA and NA subtypes. For example, A/Swine/Iowa/15/30 (H1N1) was one of the earliest swine flu viruses isolated (in 1930). The A/PR/8/34 (H1N1) virus, found in a human being in Puerto Rico, shows that viruses of the same subtype, H1N1, were circulating through the world in the early 1930s in both swine and humans.

The adjacent table presents all the different H and N subtypes so far identified and the names of the prototype viruses that contain these immunologically distinct protein subunits. Note that humans, swine, and ducks can all be infected with an influenza A virus containing the H1 protein (PR/8/34, Sw/Ia/15/30, and Dk/A1b/35/76). Similarly, human, swine, and chicken virus subtypes share

the same N1 protein. These studies demonstrate two important facts about the influenza A virus: (1) There is a good deal of variation in the HA and NA proteins, so immunologically distinct types readily arise; for example, anti-H1 antibody will not protect a human being against an H2 virus subtype. (2) These influenza A subtypes have a very broad host range; they are found in a large number of animals and cause a similar disease in each.

We are now in an excellent position to explore the reasons that influenza A has pandemic and epidemic cycles and why it is able to infect the same host several times. Recall that influenza A virus has two distribution patterns, both unusual: at irregular intervals, often years apart, a pandemic appears to begin in east Asia and sweep west, accompanied by high mortality rates; alternatively, between pandemics, less severe and more localized epidemics

Hemagglutinin and Neuraminidase Subtypes of Influenza A Virus				
	Species of origin			
Subtype	Human	Swine	Horse	Bird species
H1	PR/8/34	Sw/Ia/15/30		Dk/Alb/35/76
H2	Sing/1/57			Dk/Ger/1215/73
H3	HK/1/68	Sw/Taiwan/70	Eq/Miami/1/63	Dk/Ukr/1/63
H4				Dk/Cz/56
H5				Tern/S.A./61
H6				Ty/Mass/3740/65
H7			Eq/Prague/1/56	FPV/Dutch/27
H8				Ty/Ont/6118/68
H9				Ty/Wis/1/66
H10				Ck/Ger/N/49
H11				Dk/Eng/56
H12				Dk/Alb/60/76
H13				Gull/Md/704/77
H14				Mall/Gurjev/263/82
N1	PR/8/34	Sw/Ia/15/30		Ck/Scot/59
N2	Sing/1/57	Sw/Taiwan/70		Ty/Mass/3740/65
N3				Tern/S.A./61
N4				Ty/Ont/6118/68
N5				Sh/Austral/1/72
N6				Dk/Cz/56
N7			Eq/Prague/1/56	FPV/Dutch/27
N8			El/Miami/1/63	Dk/Ukr/1/63
N9				Dk/Mem/546/74

affect defined areas rapidly every couple of years, often infecting some of the same people previously infected with the influenza A virus.

Over the past hundred years, there have been five great pandemics, beginning in 1890, 1900, 1918, 1957, and 1968. The so-called Spanish influenza of 1918 and 1919 killed between twenty and forty million people and crippled the armed forces at the end of World War I; about 80 percent of U.S. Army deaths that year were due to influenza infec-

Influenza A Pandemics of the Past Hundred Years	
Year of origin	Subtype in circulation[1]
1890	H2N8
1900	H3N8
1918	H1N1 (Spanish flu)
1957	H2N2 (Asian flu)
1968	H3N2 (Hong Kong flu)
(1977)	H3N2 and H1N1

[1] The reintroduction of H1N1 in 1977 resulted in two viruses presently circulating in the population, but did not cause a pandemic.

tion. Because the first isolation of a human influenza A virus did not come until 1933, it was only possible to obtain viruses from later pandemics (1957 and 1968). The so-called Asian flu circulating in the 1957 pandemic was shown to be an H2N2 virus that had suddenly replaced the H1N1 virus circulating in the human population prior to that year. Similarly, a new pandemic strain that arose in 1968, the so-called Hong Kong flu, contained an H3N2 shift and quickly replaced the H2N2 virus in circulation between 1957 and 1968. Seroarcheological techniques—testing the antibodies of people who lived through these epidemics—have demonstrated that the 1890 strain was an H2N8-like virus, the 1900 strain was an H3N8-like virus, and the 1918 strain was most likely an H1N1 virus, which appeared again in 1977 and is still in circulation, along with H3N2.

Every new pandemic is accompanied, it is clear, by a change in the HA subunit on the surface of the influenza A virus particle; in some cases, there is also a change in the NA subunit. Although the population retains its lifelong immunity to the HA antigen of the strain that infected it in the past, the new pandemic subtype changes its HA so that neutralizing antibodies no longer react with epitopes A, B, C, D, and E on the new HA molecule. Somehow the new strain has changed five or more sites on its HA protein simultaneously. Because a separate mutation is required to alter each of the five epitopes, segment 4, the RNA chromosome that encodes the information for the HA protein, must have undergone at least five, and in many cases more, mutations.

We have seen in previous chapters that the probability of two independent mutations occurring is the product of the individual probability of each event. If a mutation occurs in one in a million viruses—a rare event—so as to change one epitope, then two mutations in that same virus will occur only once in a million million (10^{12}) viruses. Five independent mutations in the same virus, each one changing an epitope at A, B, C, D, and E, would occur only once in 10^{30} viruses. There are simply not that many viruses in the world. Single, independent mutations between 1900 and 1918 could not explain how the H3N8 virus was replaced by an H1N1 virus, because there were not 10^{30} influenza A viruses replicating during those eighteen years. Rather, the H3 and H1 genes must already have existed in a viral population before the great Spanish flu pandemic. H3 virus was clearly circulating in humans before 1918, but H1 antibodies were only detected in humans after the 1918 pandemic. Could an H1 strain have existed in pigs and ducks in 1918? Even if it had, how could the H1 gene move from a subtype affecting those animals into a new strain so virulent for humans?

Sialic acid receptors are common in all species, so it is not surprising that influenza A virus can enter the cells of many different animals and replicate, although poorly (until it accumulates genetic changes to help it reproduce in the new host species). Suppose the human H3N8 virus, in circulation before 1918, infected a duck carrying an H1N1 strain. If the human H3N8 subtype and the duck H1N1 subtype both infected the same cell, an interesting problem would arise.

The human virus brings in eight RNA chromosomes—let's call them segments 1hu, 2hu, 3hu, 4hu, 5hu, 6hu, 7hu, and 8hu. The duck virus brings to the cell eight RNA chromosomes—1dk,

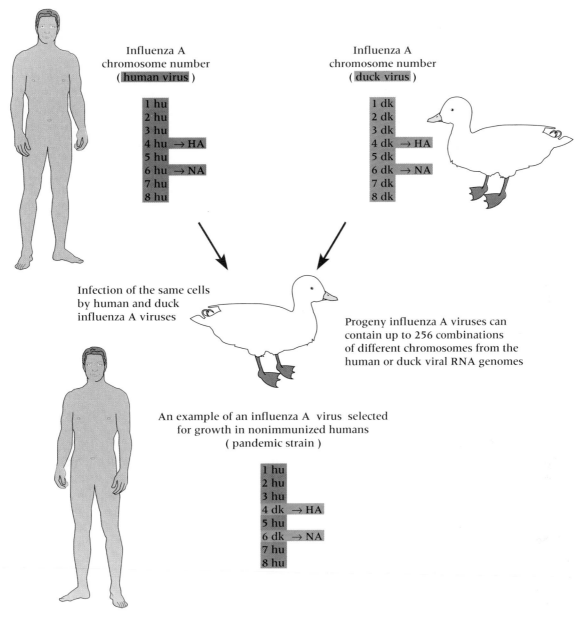

Infection of the same cells by human and duck influenza A viruses

Progeny influenza A viruses can contain up to 256 combinations of different chromosomes from the human or duck viral RNA genomes

An example of an influenza A virus selected for growth in nonimmunized humans (pandemic strain)

Infection of one cell with both human and duck influenza A viruses can result in a reassortment of viral chromosomes, followed by selection for a virulent progeny virus not previously present in the human population. Animal reservoirs, which harbor influenza A virus with distinct chromosomes producing a wide variety of HA and NA antigenic subunits, are the source for new viral genes.

The Influenza A Virus

2dk, 3dk, 4dk, 5dk, 6dk, 7dk, and 8dk. One of these agents is an influenza A virus that has replicated in human hosts over the years, and the other is an influenza A virus that has replicated in ducks over the years, so chromosomes 1hu and 1dk are related but are not identical. The dk genes produce a strain that grows best in ducks; similarly, the hu genes make a virus that replicates well in the human lung. When both viruses find themselves in the same cell in the lung of a duck, they make all ten respective human and duck viral proteins (twenty in all) and then go on to replicate the eight respective chromosomes that constitute the influenza A viral genome from human and duck. As the newly copied chromosomes are packaged into virus particles, the eight hu RNA segments and the eight dk RNA segments (sixteen in all) may assort independently and be placed indiscriminately into different virus particles.

Imagine that a virus is created, for example, that contains segments 1hu, 2hu, 3hu, 4dk, 5hu, 6dk, 7hu, and 8hu. Eight of the genes (six of the chromosomes) in this virus are human (H3N8), and two of the genes are duck (H1N1). (Segment 4 of the viral genome codes for HA; segment 6 codes for NA—see page 158.) Because the majority of genes are derived from a human strain, the virus will replicate pretty well in humans. In addition, however, the new strain is H1N1 in a world where the potential human hosts have previously been infected with an H3N8 subtype, but have never before encountered an H1N1 strain. This virus is thus selected for its ability to replicate in a population with no immunity to it. The emergence of this subtype is like the exposure of the Mexican Amerindians to smallpox. A pandemic ensues.

This hypothesis is interesting, but is it correct? One can deliberately reproduce this experiment in cell culture. When two different influenza A viruses are mixed to infect the same cells in culture, RNA segments can be tested for, identified, and observed to assort randomly into the progeny virus particles in various combinations. Because two different viruses each denote eight subunits, 256 combinations

of RNA segments are theoretically possible. These experiments demonstrate that (1) different influenza A strains can infect the same cell, and (2) the RNA chromosomes of these subtypes can be sorted independently into different particles.

Does this happen in nature? Several studies suggest that it does. In the seal influenza A strain A/Seal/Mass/1/80 (H7N7), all eight RNA segments are closely related in nucleotide sequence to several different avian influenza A subtypes. It has been suggested that seals were repeatedly infected with mixtures of avian influenza A viruses by contact with feces from infected birds. The reassortment of avian RNA segments then selected a virus that causes high mortality in seals. (Curiously, this influenza A virus is pathogenic for humans as well; accidentally transmitted to humans working with seals and to laboratory workers growing this strain, it caused conjunctivitis, an eye infection. This broad host range is quite unpredictable. The strain is quite pathogenic in some monkeys, for example, and replicates in pigs but produces no symptoms.)

Genetic reassortment of influenza A viruses has been observed in people as well. The subtypes now in circulation, H3N2 and H1N1, were shown to produce reassortant viruses in humans during an epidemic in 1978 and 1979. Because the reassorted chromosomes were not observed in further epidemics, we can conclude that these viruses were not selected for under natural circumstances. Additional evidence for the reassortment hypothesis comes from a detailed study of the pandemic strain of 1968. At that time, a new subtype, H3N2, replaced the H2N2 virus then in circulation. It turned out that seven RNA subunits of the H3N2 virus were closely related to the old H2N2 virus—that is, their nucleotide sequences suggested that almost all of the RNA segments of the 1968 virus derived from the previous H2N2 agent. The one exception was the gene for the hemagglutination antigen, now H3, which was clearly different from the former H2 antigen gene. Interestingly, the H3 HA gene was 98 percent identical in nucleotide sequence with that of an avian influenza A subtype isolated in 1980. The avian strain may have donated its HA gene to the 1968 pandemic influenza A virus of humans.

Several facts, then, explain how pandemics of influenza A virus arise. Human and animal influenza A subtypes occasionally initiate mixed infections, from which may emerge a large number of possible strains containing different combinations of genetic information from the human and animal subtypes. A few of these combinations are selected for their ability to present a new HA antigen to hosts previously immunized with a different HA subtype. New pandemic strains are possible as long as there is a source of HA genes not previously encountered by living human hosts. The evolutionary process has, in effect, been speeded up by the influenza A virus's ability to reassort genes rapidly and by the existence of animal reservoirs that can be tapped for new genetic combinations. As a result, influenza A virus has many opportunities to survive and flourish throughout the world.

These antigenic shifts may explain pandemics, but not the constant local epidemics of influenza A that occur between pandemic years. Once a new virus subtype results in high rates of infection

around the globe, the next years see that subtype reinfect some of the individuals previously infected with that same strain. That is, the viruses within a single subtype (H1N1, for example, or H3N2) also seem to change with time.

A clear way to demonstrate this is to examine the HA antigen from influenza A viruses derived from a given subtype—say, H3N2—between pandemic years—say, 1968 through 1986. During this time, nine different virus samples were obtained from local epidemics all over the world—Hong Kong (8/68), England (42/72), Port Chalmers (1/73), Victoria (3/75), Texas (1/77), Bangkok (1/79), the Philippines (2/82), Mississippi (1/85), and Leningrad (360/86). To determine how closely related the HA proteins of these nine viruses are, the hemagglutination property of the HA protein was exploited.

Recall that when the HA protein is mixed with red blood cells, it binds to sialic acid on the cell surfaces, linking them together so that they clump in a test tube. When the antibody binds to site A, B, C, D, or E of the HA subunit, it blocks the HA from binding to sialic acid and inhibits hemagglutination. Hemagglutination inhibition was tested with antibody made against each of the nine influenza H3N2 viruses. If antibody directed against virus 1 also inhibits hemagglutination by the HA of virus 2, the two viruses are related.

It is possible, moreover, to quantitate this relationship. For example, if a volume of antibody 1, made against a specific virus sample 1, comprises 320 units of antibody that inhibits hemagglutination, it may contain only 80 units of antibody that inhibits hemagglutination by the HA of virus sample 2. The number of antibody units that block hemagglutination measures the degree of relatedness between two HA proteins, whatever form that degree of relatedness takes (for example, the vulnerability of more sites on the HA proteins to antibody binding, or the tighter binding of antibody on both proteins to a particular site—like A—that blocks hemagglutination). The useful thing about this test is that the strength of hemagglutination

The consequences of antigenic shift, 1918.

inhibition correlates with the efficiency of neutralizing antibodies. If the same antibody inhibits hemagglutination by two viruses, it is likely that prior infection with one virus will protect against infection with the second.

The results of such a test, performed on several influenza A subtypes, are presented in the table below. Antibodies were made against each of five viruses, and the amount of each antibody that successfully inhibited hemagglutination is given for each of the nine viruses tested. For example, antibody HK/8/68 works best against the virus whose HA induced the antibody response—A/Hong Kong/8/68; 320 units of this antibody reacted successfully against this viral strain. The antibody reaction table shows that A/England, A/Port Chalmers, and A/Victoria are all related to Hong Kong; A/Bangkok is virtually identical with Hong Kong; but A/Texas, A/Philippines, A/Mississippi, and A/Leningrad are distant from the A/Hong Kong virus. In general, the antibody always works best against the virus it was made against in an infected host (note the diagonal pattern, underlined, in the table).

With some exceptions, moreover, the antibodies recognize as closest the viruses that were in circulation at about the same time as was the isolate under study. For example, antibodies directed against A/England/42/72 react equally well with viruses isolated in 1968 and 1972, about half as well with viruses isolated in 1973 and 1975, about 10 to 25 percent as well with viruses isolated in 1977 and 1979, and not at all with viruses isolated in 1982 through 1986. Clearly, these viruses are changing with time. Even though prompted by an influenza A H3N2 virus, antibody made by someone infected with A/Hong Kong/8/68 may not protect you from A/Leningrad/360/86.

What happened here is a series of mutations in the HA gene. Millions of influenza A viruses were produced by variants carrying mutations in both HA sites A and B—mutations that arose, not simul-

	Hemagglutination Inhibition of H3N2 Influenza A Viruses Isolated Between 1968 and 1986				
	Antibody units that inhibit hemagglutination				
Virus	HK/8/68	E/42/72	PC/1/73	Vic/3/75	Tex/1/77
A/Hong Kong/8/68	320	320	0	0	0
A/England/42/72	80	320	80	40	0
A/Port Chalmers/1/73	80	160	320	80	40
A/Victoria/3/75	80	160	320	640	160
A/Texas/1/77	0	40	160	160	1280
A/Bangkok/1/79	320	80	320	320	1280
A/Philippines/2/82	0	0	0	0	80
A/Mississippi/1/85	0	0	80	40	160
A/Leningrad/360/86	0	0	0	0	80

Experimental results adapted from M. W. Harmon and A. P. Kendal, American Public Health Association: 1989.

taneously, but sequentially in the same virus. In one in a hundred thousand or a million viruses, a single mutation changed the A epitope on the HA protein. Then, when there were a million A-mutant viruses in circulation, a mutation occurred, altering site B, at a similar frequency. One virus now had both A and B epitope variations, and that strain reacts poorly with the antibody of a previously immunized host. This host, disarmed, may become reinfected with the new strain of influenza A virus. The alteration of one or two HA sites has been called antigenic drift to contrast it with antigenic shift—the complete change of the viral RNA segment coding for the HA protein.

Different people will respond differently to such influenza A mutants. Some individuals make plenty of highly reactive antibodies that are able to neutralize influenza virus at site C, D, or E alone. Other people produce low levels of antibody, or antibodies that do not seem to bind to a particular epitope very well. Part of our genetic endowment, what makes each of us an individual, is how well we respond to a given virus infection. Not only does the circulating virus have a great deal of genetic variability, but the host's immune response is variable as well. Some people, then, can be reinfected with influenza A virus that has mutated at only a single epitope: A, B, C, D, or E.

One more question remains to be answered about antigenic drift in influenza A virus. All living things change their nucleotide sequences with time. During the duplication of genetic information, mistakes are made and nucleotide sequences are altered. In fact, such mutations are essential for continued life. Over millions of years, species must change to be able to adapt to a changing environment. This new environment selects, from among the altered examples of life, the ones most fit to survive, which move on to the future. If this is true for every living thing, why does influenza A virus seem to change faster than most? Why can this agent alone reinfect a human host who retains lifelong immunity to other viruses? Why don't all viruses adapt like influenza A virus? Although we do not have all the answers to these questions, two

Paris grippé, one of a series of lithographs of Parisian life by Honoré Daumier (1808–1879), documents the familiar results of influenza A antigenic drift.

insights should partially satisfy us. Both reflect the fact that influenza A virus has an RNA genome, whereas almost all other living organisms store their genetic information as DNA.

A good case has been made that the earliest life forms on Earth used RNA polymers both to store information in the nucleotide sequence and to catalyze chemical reactions. Studies over the past few years have shown that, under the right conditions, RNA polymers can replicate and even cut themselves into pieces of defined sizes. RNA is a very reactive polymer, but this very property makes it poorly suited to store information. If our goal were to store information stably, whether on a computer tape or in a polymer, we would not want all sorts of chemical reactions to occur over the length of our tape that might alter the information stored there. For a variety of chemical reasons, DNA is much less reactive than RNA and therefore is better suited to serve as an inert storage molecule. Most living organisms, accordingly, use RNA to carry the messages for protein production, but not to store their genomic information.

Since RNA has greater reactivity than DNA, RNA viruses evolve more rapidly than do DNA viruses. The source of RNA's higher mutation rate can be seen by comparing the duplication of RNA with that of DNA. As the enzyme RNA replicase copies one strand of RNA into a complementary polymer, it averages one mistake per 10,000 (10^4) nucleotides copied. As DNA polymerase copies DNA into its complementary strand, it makes about one mistake in every million to ten million nucleotides (10^6 to 10^7). Part of this difference in accuracy is due to the reactivity of RNA and to the variable fidelity of the duplicating enzymes.

Another part reflects the fact that DNA polymerases have evolved the ability to correct mistakes in their synthesizing process. In DNA, A on one strand pairs with T and G pairs with C. If, in copying DNA, G were paired with T by mistake, DNA polymerase could register the error and correct it. In effect, DNA polymerases, but not RNA polymerases, have proofreading abilities. RNA genomes, therefore, retain more mistakes in their copies than do DNA genomes.

There is an additional reason that influenza A viruses evolve rapidly. All viruses, as we have seen, go through a very large number of duplication cycles in a short time. A single viral RNA genome may reproduce ten thousand copies of itself in six hours, while the generation time of the host for an influenza A virus is measured in years—for humans, perhaps a quarter century. Not only is the error frequency in the original RNA genome high, but the large number of generations yearly causes the progeny viruses to diverge rapidly. It has been estimated that 0.03 to 2 percent of the nucleotides in the genome of an RNA virus are altered every year. Some RNA viruses—those whose polymer-ases have high error frequencies—are evolving much faster than any other living organisms.

When mutations arise in a virus, they occur randomly across the entire length of the genome. We are able to recognize three basic classes of mutations: (1) neutral mutations, which change the nucleotide sequence but give the virus no apparent advantage or disadvantage over the parent; (2) detrimental mutations, which select against the virus; and (3) advantageous mutations, which over time produce the majority of viruses in a population. Advantageous mutations are the rarest. A virus can tolerate a very high mutation rate only if the level of detrimental mutations is not too high, so viruses that can maximize neutral or advantageous mutations are better able to tolerate high rates of mutation randomly across the genome. Influenza A virus appears to have many of the elements necessary to undergo antigenic drift—a high mutation rate, many generations per year, and plasticity of structure.

In response to the obstacle posed by the immune systems of its hosts, influenza A virus has developed two fundamental modes of change: antigenic shift and antigenic drift. The question remains: How do we fight back against a virus programmed for change? This question led to the development of the swine flu vaccine in the United States in 1976—and the subsequent public health fiasco. It is a story worth repeating.

The Swine Flu Vaccine

By 1975, scientists had a good idea of how twentieth-century influenza A pandemics came about: beginning in 1900, 1918, 1957, and 1968, viruses with different HA subunits (H3, H1, H2, and H3 again) had swept across the world, attacking unimmunized hosts. Many proposed that we should constantly be on the lookout for influenza A strains with new HA proteins, since such a virus might originate a new pandemic. A vaccine made against

Die Familie (The Family) by Egon Schiele (1917). This famous painting by the Austrian expressionist contains a self-portrait, although he used models for the wife and child. The following year, Schiele's own wife, Edith, was in the sixth month of her first pregnancy when she contracted the Spanish flu and died within forty-eight hours. Four days later, the same pandemic strain of influenza A virus killed the artist. He was twenty-eight years old.

this new subtype could be administered to the world's population before the disease spread. It appeared that we knew both what to look for and what to do about it.

These concepts were under discussion when, in mid-January 1976, a large number of respiratory illnesses were reported among Army recruits at Fort Dix, New Jersey. On January 29 and 30, nineteen throat washings from patients were sent to the virus laboratory at the New Jersey Department of Health, and from these samples, eleven influenza A virus isolates were obtained and typed by HA inhibition tests (using the antibodies in the table on page 170). Most were identified as A Victoria or A Port Chalmers, which were circulating at the time, but two samples could not be identified by antibodies in the test group and five isolates gave

unclear results. The state health officials sent these seven samples to the federal Centers for Disease Control in Atlanta, Georgia. While testing was in progress, one of the Fort Dix recruits left his sickbed to participate in a forced five-mile march. Following the march on February 4, Private David Lewis collapsed and died. On February 5, the CDC confirmed five New Jersey samples as influenza A Victoria, but two did not type as any known virus in circulation. By February 10, New Jersey health officials had sent two more unidentifiable isolates of the virus from Fort Dix to the CDC; one of these was from David Lewis. On February 12, the four isolates from Fort Dix were typed as influenza A virus with swine flu type hemagglutinin, H1.

A virus with the H1 antigen had not been seen in the human population since 1918, when the

H1N1 subtype caused the Spanish flu at the end of World War I, killing tens of millions of Europeans. The rapid and deadly pattern of respiratory disease from this strain was the worst of the century. In major cities like Vienna, the epidemic was so bad that streetcars were converted into public hearses.

Health officials in the United States, aware of H1 influenza A's terrible history, were alarmed. By February 13, 1976, the H1 swine flu virus isolates were reconfirmed at the CDC using new tests, and the next day an emergency meeting of scientists and researchers was held. For the first time, the possibility of making a swine flu vaccine was raised. On February 19, a CDC press conference announced the isolation of a swine flu strain at Fort Dix. The prepared remarks made no reference to the 1918 pandemic, but reporters asking the right questions brought up its relation to the present virus isolates. By February 20, there was wide-spread coverage of this news story.

During the week of February 20 to 27, a subsidiary of the Food and Drug Administration delivered the Fort Dix virus to four vaccine-manufacturing companies. Additional swine flu isolates turned up in Minnesota, Wisconsin, Pennsylvania, Virginia, and Mississippi, but all these cases appeared to involve people who had had extensive contact with pigs. Whether the virus spread from human to human and its degree of virulence remained unclear. By March 9, a survey of recruits at Fort Dix showed that as many as five hundred might have been exposed to the virus strain, although there had been only one death. By the 12th, several groups were recommending that swine flu be included in a vaccine with A Victoria and B Hong Kong (influenza B virus is a distinct agent). Some government health officials agreed to immunize members of the armed forces, and others wanted nationwide immunization.

On March 24, President Gerald R. Ford met with a blue-ribbon panel of experts at the White House, and later went before the television cameras to recommend a vaccination program for all Americans. In mid-April, the president signed a bill authorizing funds for this program.

Full vials of swine flu vaccine tumble off the assembly line at an American pharmaceutical company, October 1976.

Throughout May and June, pharmaceutical companies began producing swine flu vaccine with the goal of 196 million doses (reflecting the population of the United States at the time) by November 1—just before the expected influenza season. In July, insurance carriers for each of these companies, skittish about financial risk, cancelled liability coverage for swine flu vaccine. Many scientists, including Albert Sabin, urged that the vaccine be stockpiled pending a new outbreak; others discouraged parallels between 1918 and 1976. At this juncture, the drug companies informed the Department of Health, Education, and Welfare that they planned to stop manufacturing vaccine because they were unable to get liability insurance.

On August 2 there was a sudden outbreak of a mysterious respiratory disease in Philadelphia. Prominent among the possible causes discussed by health officials and in the newspapers was swine flu virus. The CDC, which named the syndrome Legionnaires' disease (it was first observed at an American Legion convention), began a search for the causative agent—which, in the end, turned out not to be swine flu virus. With government officials speculating that swine flu might account for the deaths in Philadelphia, however, Congress passed a vaccine liability-protection bill that the president signed on August 12. A poll in late August indicated that 93 percent of Americans were aware of the immunization program and 52 percent wanted vaccination. The first swine flu shots were given at the Indiana State Fair on October 1, 1976.

Ten days later three elderly individuals, each of whom had received a swine flu shot at the same clinic in Pittsburgh, died suddenly. Pittsburgh health officials closed down the clinic, and nine other states followed suit. By October 14, more than a dozen persons from nine states had died shortly after receiving the shot. The president and his family were immunized on television; the government pressed on with its program. Two late-autumn reports of patients with antibodies to H1 swine flu were highly publicized and resulted in increased rates of vaccination, but neither proved, on examination, to demonstrate human-to-human transmission.

As the numbers of immunized people climbed, fifty-four cases of the Guillain-Barré syndrome (a neurological disorder) were reported, thirty of which occurred in individuals who had received a flu shot within the previous month. The lack of real evidence for a swine flu pandemic, the association of various deaths or complications with flu shots, and the rising number of voices in favor of a more cautious attitude all led the government to call, on December 16, for a one-month suspension of the immunization program.

In January 1977 a new administration was sworn in and a new Secretary of Health, Education, and Welfare was appointed. By February 4, 104 damage claims citing swine flu immunizations, totaling $11 million, had been filed. The influenza outbreak that winter turned out to be A Victoria; H1 swine flu was only present in pigs. By February 8, vaccinations were resumed—against A Victoria, A Port Chalmers, and B Hong Kong strains. By March, the targets for the next year's vaccine did not even include the H1N1 swine flu subtype.

What can we learn from this story—which, remarkably, took place over the course of one short year? The fundamental science that told us what to look for was sound, but it somehow resulted in overreaction. Election-year politics and the desire to protect the American public played their roles. Every time a few facts suggested caution, something arose to reinforce fear—for instance, the quite irrelevant episode of Legionnaires' disease. Although evidence mounted that the incidence of swine flu in the United States was the result of pig-to-human contact—and that, moreover, the virus strain caused little or no disease in people and in any case was not pandemic—the momentum of a nationwide vaccination program could be stopped only by questions of liability, not of science. It still remains unclear whether the swine flu vaccine caused sudden deaths or Guillain-Barré syndrome, and the strong counterresponse to the vaccine program was probably also excessive.

Science today is carried out by debate in the academic literature and at professional meetings. Results, opinions, and conclusions often differ; with time and despite false starts, hard-won facts and new directions emerge. Public health policy, perhaps by necessity, encompasses more components than science alone. We must learn to use the system we have to make better and wiser decisions.

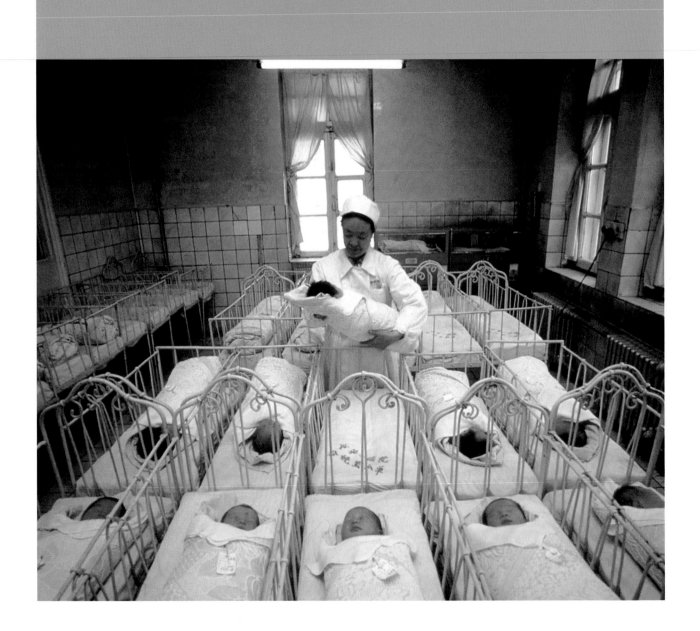

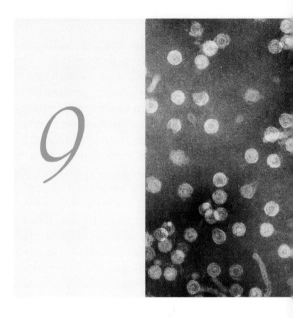

The Hepatitis B Virus

We have a habit in writing articles published in scientific journals to make the work as finished as possible, to cover up all the tracks, to not worry about the blind alleys or to describe how you had the wrong idea first, and so on. So there isn't any place to publish, in a dignified manner, what you actually did in order to get to do the work, although there has been, in these days, some interest in this kind of thing.

Richard Feynman, 1966

Hepatitis is a general term describing an infection or inflammation of the liver. At least five different viruses have been associated with hepatitis—Epstein-Barr virus, cytomegalovirus, and the hepatitis viruses A, B, and C—but it is the last three that most commonly cause liver disease. The nature and properties of these three agents are quite different, and while the symptoms they cause have some common features, the long-term effects can be disparate.

Left: Newborns in a Chinese nursery. For centuries, complex patterns of viral transmission put infants like these at high risk for chronic hepatitis B infection and the increased incidence of liver cancer it entails. But epidemiological data and biomedical technology have transformed them, within this decade, to targets of intervention rather than disease.

Above: EM of the hepatitis B virus.

Hepatitis A and C

Hepatitis A virus, a picornavirus with a structure and pattern of replication similar to those of poliovirus, is most commonly acquired from contaminated food and water. After exposure, there is a variable incubation period (fifteen to forty days) during which the virus replicates in liver tissue but does not cause overt pathology. Shortly before the onset of symptoms, virus is detectable in the blood and feces; as with poliovirus, transmission is by the oral-fecal route. Fever, malaise, and nausea are followed by jaundice, a yellow discoloration of the skin and conjunctival tissue caused by the deposition of several bile pigments into the skin and eyes as a result of the liver's failure to remove these products from the blood. The symptoms gradually reverse with time, and a vigorous immune response clears the virus and provides long-term immunity. The incidence of hepatitis A in the U.S. population is about ten cases per hundred thousand, and the geographical distribution reflects the fact that poor sanitation favors disease. Many hepatitis A infections occur with no symptoms, and a large percentage of individuals have antibody against this virus. For this reason, injection of antibody pooled from many people (immune serum globulin, or gamma globulin) is an effective prophylaxis. Vaccines—both killed and attenuated—continue to be field-tested.

The hepatitis C virus was only discovered in 1989, although its existence was suspected and the disease it caused was often termed hepatitis non-A non-B, to indicate hepatitis not caused by the A or B viruses. A group of scientists from Chiron Corporation, a small genetic engineering company in California, along with the Centers for Disease Control, applied the tools of molecular biology to find this mystery virus. Starting with a batch of plasma (blood proteins) that could transmit liver disease to chimpanzees but that did not contain the hepatitis A or B viruses, they concentrated a putative virus and extracted its nucleic acids. They next made a DNA copy of these nucleic acids and placed this putative gene from the serum virus into a foreign host, the lambda bacteriophage described in Chapter 2. With the correct signals (promoters of transcription), the lambda bacteriophage made a new protein encoded from the DNA copy of the serum virus.

Next, antibodies were obtained from people who had had hepatitis transmitted by a transfusion of blood or exposure to blood products but whose serum contained no antibodies to hepatitis A or B

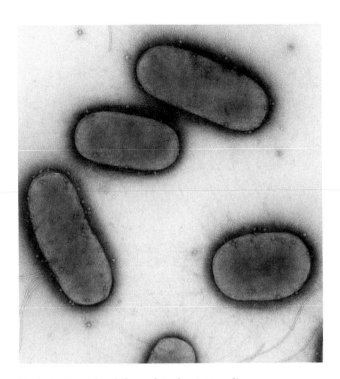

Bacteriophage lambda and its host, *E. coli.*

viruses; these patients are classified as having had non-A non-B disease. Their antibodies were tested to determine whether they would bind to the protein made by the lambda bacteriophage replicating in *E. coli* and expressing the foreign protein coded by the putative non-A non-B virus. Remarkably, the research strategy was successful, and a single gene from the new virus, called hepatitis C virus, was eventually isolated and its nucleotide sequence determined.

In the same journal issue announcing the identification of this virus, the research group reported an assay for hepatitis C antibodies in the blood of patients, this time using hepatitis C virus antigen expressed in yeast. They had observed that six of seven human blood-protein samples that could transmit hepatitis non-A non-B disease contained antibodies to hepatitis C virus, indicating that those six people had been exposed to hepatitis

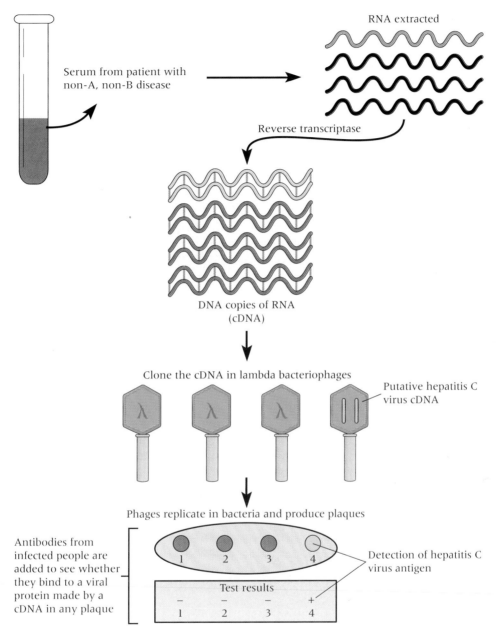

Scheme for cloning a hepatitis C virus cDNA (c is for copy) and testing for its antigen. Each cDNA molecule was packaged into a different lambda bacteriophage. Each virus yields one plaque—a clone of progeny viruses—and some plaques express the putative hepatitis C antigen. Cloning of this hepatitis C cDNA in lambda permitted the propagation of millions of copies of the cDNA in a harmless bacteriophage.

C. Of ten recent blood-transfusion recipients in the United States who had subsequently developed hepatitis non-A non-B, all made antibodies to this antigen. Further, about 80 percent of the chronic hepatitis non-A non-B post-transfusion patients in Japan and Italy had antibodies to the new protein, and about 58 percent of the non-A non-B patients in the United States were antibody-positive. Clearly, hepatitis C virus is a major cause of non-A non-B disease worldwide. The entire virus and its genome were rapidly isolated and the complete nucleotide sequence of its RNA determined.

Hepatitis C virus is a togavirus, related to the yellow fever virus (flaviviruses)—the first human disease agent shown to possess the properties of a virus (see Chapter 1). It is already clear that hepatitis C virus causes chronic liver disease and is probably linked to some liver cancers. Both a test for the hepatitis C virus in the blood supply and a possible vaccine will follow rapidly from these observations. Clearly, the tools for understanding the molecular biology of viruses, developed in the 1940s to 1960s (see Chapter 2), and their applications in DNA isolation and cloning techniques introduced in the fol-

A Comparison of Three Viral Agents Causing Liver Disease in Humans			
Properties	**Hepatitis A**	**Hepatitis B**	**Hepatitis C**
Classification	Picornavirus	Hepadnavirus	Togavirus
Virion structure			
Genome	Single-strand (+) RNA	Part double-strand (+/−) DNA	Single-strand (+) RNA
Coat	Protein icosahedron	Protein shell; lipid envelope	Protein icosahedron; lipid envelope
Virion enzymes	None	DNA polymerase; reverse transcriptase	None
Transmission	Oral-fecal route	Blood and blood products; shared needles; venereal and maternal transmission; saliva	Blood and blood products
Common name of disease	Infectious hepatitis	Serum hepatitis	Formerly, hepatitis non-A non-B
Pathology	15–40 days incubation period; acute disease only	60–160 days incubation period; acute disease and chronic disease (2–15 percent of cases) both occur	Incubation period unclear; acute and chronic diseases both occur

lowing decade have resulted in novel and important discoveries in the 1980s. Bacteriophages (like lambda), in their relative simplicity, continue to contribute significantly to the study of more complex viruses.

Hepatitis B

The hepatitis B virus is a different agent. It belongs to the hepadnavirus group, which also includes viruses of woodchucks, ducks, tree and ground squirrels, and herons, as well as some other, poorly characterized, isolates. These agents are similar but distinct viruses, containing a DNA genome that is found partially as a double-stranded helix and partially as a single strand. Closely associated is a DNA polymerase enzyme, encoded by the virus, that can polymerize nucleotides onto the short strand to complete the double helix. The viral genome is surrounded by a protein core, enclosed in turn by a lipid envelope from which protrudes lipoprotein spikes called the surface antigen (S antigen).

Hepatitis B virus used to be transmitted by infected blood and blood products. Before routine testing of the blood supply in the United States, each year there were at least thirty thousand cases of posttransfusion hepatitis, resulting in fifteen hundred to three thousand deaths. This number was reduced by 90 percent after testing for hepatitis B virus was instituted. Today, most transmission of this virus occurs through needles shared among the drug addict population, by heterosexual and homosexual venereal transmission, through close personal contact via fluids such as saliva, and from mother to newborn.

After a long incubation period, some patients have an acute disease episode with symptoms similar to those of hepatitis A virus infections. A variable percentage of these patients continue to harbor the virus for life. Chronic replication of virus in the liver can destroy the organ and also carries a very high risk of hepatocellular carcinoma. Recently, an effective vaccine was developed for hepatitis B virus and is in extensive use in some parts of the world.

Hepatitis B virus is much like a retrovirus except that its genome is DNA, not RNA. Its replication cycle is unique. Hepatitis B, moreover, is an excellent example of a virus having extensive interactions with its host, which in this case is us. This virus has learned how to replicate continuously in liver cells over the lifetime of the host, resulting in slow and progressive disease. And because this chronic, active growth of the virus often results in liver cancer, hepatitis B virus is the most common viral source of human cancer in the world, particularly in Asia and parts of Africa.

Historical Considerations

The most notable sign of liver disease, jaundice, is a clearly observable symptom, commonly mentioned in a number of ancient medical sources. The Babylonian Talmud and the description of diseases by Hippocrates both record epidemic forms of jaundice, but because any liver damage can result in jaundice, it is difficult to ascribe an agent. A clearer example of serum hepatitis was reported in 1895. About 1300 shipyard workers in Bremen, Germany, were receiving smallpox vaccine, administered by placing a drop of vaccine virus prepared from human lymph (a blood product) on the skin, which was then broken with a needle. (Did they sterilize it between uses?) Two to eight months later, 191 of these workers reported jaundice and hepatitis-like symptoms.

Throughout the twentieth century, episodic cases of hepatitis were associated with improperly sterilized syringes and needles, in diabetics taking insulin and even in patrons of tattoo parlors. By 1937, it was clear that a virus in blood or blood products could transmit this disease; in two well-documented cases, measles antibody from convalescents and yellow fever vaccines containing human serum both transmitted liver disease. About this time it also became clear that some forms of

Mass inoculations like this, early in the century, often spread serum hepatitis.

hepatitis were related to the food source (hepatitis A) and that there were likely to be at least two distinct agents.

Between 1962 and the early 1970s, several studies proved these ideas to be correct. One of the more definitive and controversial was carried out at the Willowbrook State School for the Mentally Handicapped on Long Island. At that institution, sanitation was very poor; exposure of patients to feces, saliva, and so forth was common. The rate of hepatitis in the children approached 100 percent, and newly admitted patients were virtually sure to get the disease. Researchers followed new patients carefully and even experimented with recently admitted children. Serum from one patient was injected into newly arrived children, and an oral-fecal route of transmission was demonstrated.

As studies proceeded, it became clear that two different viruses were causing hepatitis. One, spread in blood and saliva, had a longer incubation period and was called MS-2 (hepatitis B), while the other, transmitted by the oral-fecal route, had a shorter incubation period (MS-1, or hepatitis A). It was also shown that immunity to one of these viruses did not confer immunity to the other.

When these appalling conditions and the experiments came to public attention, many people were alarmed. The justification offered for such experimentation was that virtually all children introduced into the school would have gotten hepatitis during their stay anyway. Today, however, better safeguards are in place at such facilities, and human experimentation even with informed consent continues to be a complicated and much-debated issue.

Having a clear picture that two different viruses were causing hepatitis in humans and that one was transmitted in blood should have made the search for this virus relatively easy. But that is not the way life or science seems to work. The discovery of the hepatitis B virus followed a tortuous course, beginning in 1963 at the National Institutes of Health where Baruch Blumberg was studying blood samples. Blumberg was testing his hypothesis that slight differences in the amino-acid sequence of "identical" proteins found in a wide range of individuals would correlate with particular disease patterns—the incidence, say, of cancer in those individuals. These subtle genetic differences, termed polymorphisms (from the Greek for "many shapes"), are known to exist in human populations and, in some cases, have indeed been linked with functional disorders.

Blumberg's approach—which aimed at the broadest perspective on these issues—was to collect serum samples from around the world. To look for differences in these samples, he used the antibodies some people make when they get a blood transfusion and their immune system detects a different protein in the foreign blood. Individuals with hemophilia (a blood-clotting deficiency) get multiple transfusions and blood products, so they were a good source of antibodies against polymorphic proteins.

In the course of these studies, Blumberg noted that a serum sample from an Australian aborigine contained a protein antigen that reacted with an antibody in the blood of an American hemophilia patient. He called it the Australia (Au) antigen. A survey of the presence of this antigen in the blood of different groups of people showed several curious correlations: (1) the Au antigen was rare in people from North America and Europe; (2) this antigen was common in Asians and Africans from certain specific locations; and (3) individuals with leukemias, leprosy, and Down's syndrome commonly had the Au antigen.

For several years, this confusing distribution led to fruitless hypotheses. An early clue to the nature of the Au antigen was provided when one of the technicians working with human serum contracted hepatitis. After she returned to work, it was noted that her serum was Au antigen-positive; before her illness, she was Au antigen-negative. By 1968, studies by A. M. Prince in New York and K. Okochi and S. Murakami in Japan had clearly established that the Au antigen was found only in the serum of hepatitis B patients. This major research advance permitted, for the first time, a specific test for hepatitis B virus. By 1970, the virus particles had been identified in patients' serum by D. S. Dane and his colleagues; they are frequently called Dane particles.

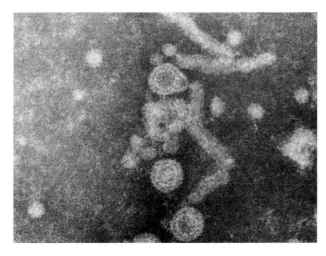

EM of hepatitis B virions (Dane particles) in the blood of a patient. The virus has a lipid envelope featuring lipoprotein spikes around a protein core containing the DNA genome.

The Structure of the Hepatitis B Virus

The hepatitis B virus is a spherical (sometimes rod-shaped) particle about 42 to 47 nm in diameter. The viral DNA is partially doubled-stranded and partially single-stranded—one polynucleotide strand is about 3400 nucleotides long, while the second varies in length from 1700 to 2800 nucleotides. Between 50 and 85 percent of the DNA, then, is double-stranded and 15 to 50 percent remains single-stranded. Under the appropriate conditions, a DNA polymerase that is encoded by a viral gene can use the long DNA strand as a template and add nucleotides to the short DNA strand, making the entire genome a double-stranded helix. This occurs shortly after virus infection of a cell.

Surrounding the DNA is a spherical shell 22 to 25 nm in diameter, made up of proteins termed the core antigens, or HBcAg (hepatitis B core antigen). The core is surrounded by a lipid envelope about 7 nm thick that derives from the host-cell membranes. In this envelope are lipoprotein spikes (S antigen, or HbsAg) that attach the virus to a liver cell. HBsAg, the protein against which neutralizing antibodies are made in the host, is the Au antigen Blumberg first discovered in 1963.

The virus encodes four known genes and proteins: the HBsAg gene; the polymerase (P) gene; the HBcAg (core) gene; and a fourth gene termed X that makes a protein not found in the virus particle. The viral gene organization is unusual in that several of the viral-encoded genes use the same nucleotides to produce different proteins.

The sequence of DNA nucleotides determines the amino-acid sequence of the protein encoded. Three nucleotides, to be precise, determine the position of one amino acid; the sequence ATG, for example, encodes the amino acid methionine, and AAA encodes lysine. In a sample DNA strand—ATGAAA—we could read the sequence ATG-AAA (methionine-lysine), but we could also read A-TGA-AA or AT-GAA-A, stipulating other amino

20 × 20–200 nm (filament)

15–25 nm (sphere)

Incomplete viruses
containing HBsAg

42 nm

HBsAg-
bearing
particle
in blood
(Dane
particle)

Virion

Weak
detergent

28 nm

Virion core
with
HBcAg

Virion core

Strong
detergent

Soluble core
antigen
released
from virion

Strong
detergent

Viral DNA

acids for the protein sequence. Where we begin on the linear DNA molecule determines our so-called reading frame for mRNA translation, and because the nucleotides encode amino acids in a three-to-one ratio, any nucleotide sequence has three possible reading frames. Genes, then, do not literally overlap, but are functionally discriminated by the reading frame.

In most viral or cellular chromosomes, a single reading frame is used to encode proteins. In the hepatitis B virus, however, the same nucleotide sequence is read in two different frames to produce two different proteins (pol and HBsAg). The use of multiple reading frames, which we have seen before in the retroviruses (Chapters 6 and 7), minimizes the amount of DNA required to encode proteins. But it has its hazards for the virus: should a mutation alter a single nucleotide, up to three viral proteins could change simultaneously.

The polymerase gene comprises the entire S-gene sequence, a part of the X gene, and a part of the core gene. The S gene actually produces three proteins, using three different start sites for the

Schematic drawing of the hepatitis B viral particle, antigens, and DNA. Small spherical and long filamentous incomplete virus particles are observed; both are composed of the envelope and the surface (S) antigen. Weak detergent dissolves the lipid membrane and exposes the nucleoprotein core; strong detergent releases the partially single-stranded viral DNA and a soluble core antigen.

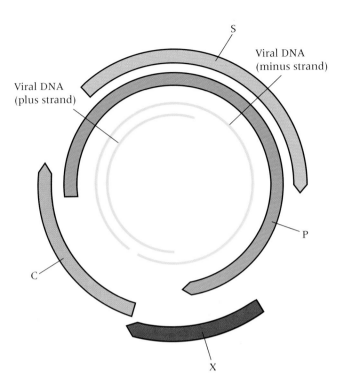

Viral DNA (plus strand)

S

Viral DNA (minus strand)

P

C

X

Map of the hepatitis B virus genome, showing the nucleotide sequences that encode the S antigen, the polymerase (P), the core protein (C), and the X protein. Different reading frames for the DNA discriminate different amino-acid sequences.

mRNAs. The P gene makes a large protein composed of a DNA polymerase and a protein that acts as an anchor for starting DNA replication; the long DNA strand, then, has this protein attached to its first nucleotide. The X gene product, a small protein of 154 amino acids, can enhance the level of mRNA synthesis of some viral and cellular genes, but its precise function remains unclear.

The Hepatitis B Virus Replication Cycle

The hepatitis B virus, attached to a liver cell via the HBsAg, enters the cell by membrane fusion or through endocytic vesicles; the exact process is unknown. The virion core enzymes convert the partially single-stranded DNA to a complete, double-stranded DNA helix. The entire genome is then transcribed in the cell nucleus, producing an RNA that is a full-length copy of the viral DNA, as well as mRNAs of shorter length. The longest RNA transcript (3400 nucleotides) acts as a template for a reverse transcriptase that is also the DNA polymerase. The reverse transcriptase copies the RNA template into a complementary DNA strand. A portion of the P protein initiates DNA synthesis on this RNA template, which is degraded as DNA polymerization proceeds, leaving a single DNA strand. The second strand of DNA (initiated by an RNA primer) is synthesized over 50 to 85 percent of the first DNA strand before sensing a stop signal.

While this replicative phase is proceeding, the hepatitis B mRNAs for core, polymerase, X, and S proteins are being translated in the cytoplasm. The full-length RNA is assembled into cores along with polymerase, and DNA replication occurs in cores within the cytoplasm (recall that DNA transcription occurs in the nucleus). The S proteins are inserted into the host cell's endoplasmic reticulum, and the

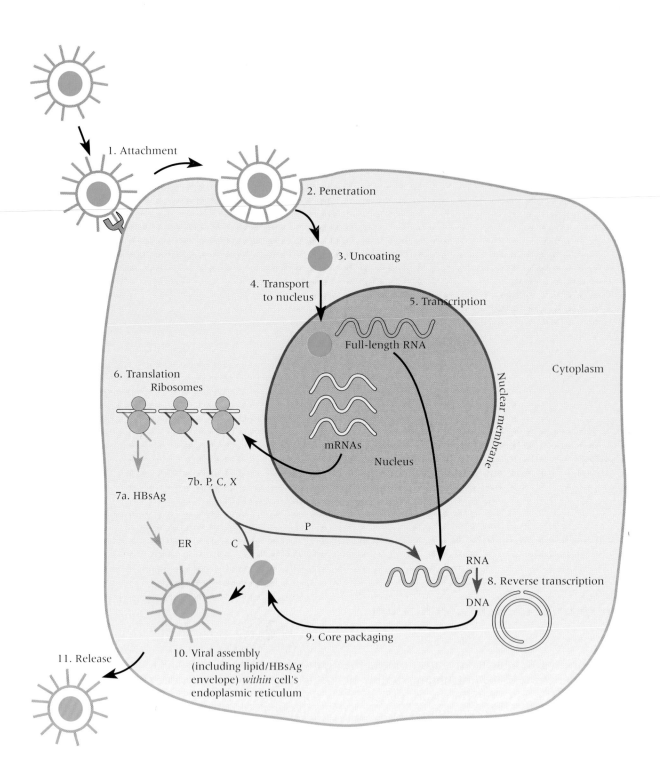

1. Attachment

2. Penetration

3. Uncoating

4. Transport to nucleus

5. Transcription

Full-length RNA

Cytoplasm

6. Translation
Ribosomes

mRNAs

Nucleus

Nuclear membrane

7b. P, C, X

7a. HBsAg

P

ER   C

RNA

8. Reverse transcription

DNA

9. Core packaging

11. Release

10. Viral assembly
(including lipid/HBsAg
envelope) *within* cell's
endoplasmic reticulum

cores appear to acquire their HBsAg envelopes within the cell. Progeny viruses then escape, shed into the extracellular fluids as mature virions.

Virus replication in liver cells is very efficient. Concentrations in the serum of infected patients can exceed one billion (10^9) virus particles per milliliter. The same serum can have up to a trillion (10^{12}) HBsAg or S proteins per milliliter, packaged in spherical lipid or membrane vesicles that do not contain the DNA core. Electron micrographs of the virus in serum from patients often show small HBsAg particles and complete, larger Dane particles.

This replicative cycle is unique to the hepadnaviruses. It is clearly related to that of the retroviruses, using a reverse transcriptase step in the duplication of the viral genome. It appears, however, that unlike the retroviruses, which have RNA genomes in the virus and DNA genomes in the infected cell, the hepadnaviruses have viral DNA genomes. One of the consequences of this reversal is that, in contrast to the retroviruses, the viral genome does not integrate into the host chromosome; more specifically, it is not required to integrate into the host-cell DNA for a successful replicative cycle. The hepadnaviruses do resemble retroviruses in other ways—the organization of the DNA polymerase gene with RNaseH, the produc-

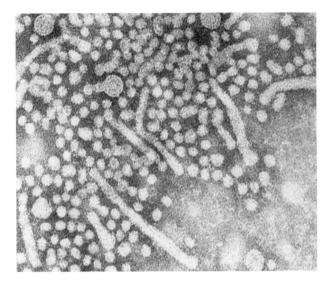

EM of the blood of a hepatitis B patient, showing virions (Dane particles) and both long filamentous and small spherical surface-antigen particles. Billions of these assorted particles may be present in each milliliter of blood.

tion of DNA from RNA within a core particle, and the coding regions that use multiple reading frames. In effect, the hepadnaviruses are specialized examples of retroviruses, or reciprocal retroviruses.

Schematic drawing of hepatitis B virus replication in a liver cell. After adsorption via its S-antigen receptor, penetration, and uncoating, the virion core moves into the nucleus, where its DNA is transcribed into mRNAs for viral proteins, as well as full-length mRNA copies of the entire genome. The shorter mRNAs move to the ribosomes and translate S, P, C, and X proteins; the HBsAg protein is localized in the endoplasmic reticulum (ER) of the cell. The full-length RNAs are packaged within the cytoplasm by the P and C proteins. An RNA-to-DNA step (reverse transcription) is carried out by P, synthesizing a double-stranded DNA genome and degrading the RNA (via RNaseH). The DNA cores move to the ER and are enveloped by lipid and the S antigen, then exit from the cell.

The Pathology of Hepatitis B Infections

The responses in individuals undergoing a primary (first-exposure) infection with hepatitis B virus vary a great deal. Mild or subclinical infections are the most common, but some patients demonstrate all the symptoms of acute hepatitis (fever, malaise, nausea, anorexia, jaundice, and so on). The severity of disease may depend upon the age of the patient (infants and young children show few symptoms), the infecting dose of virus (high doses are associated with more acute disease and shorter latent periods), and the route of entry.

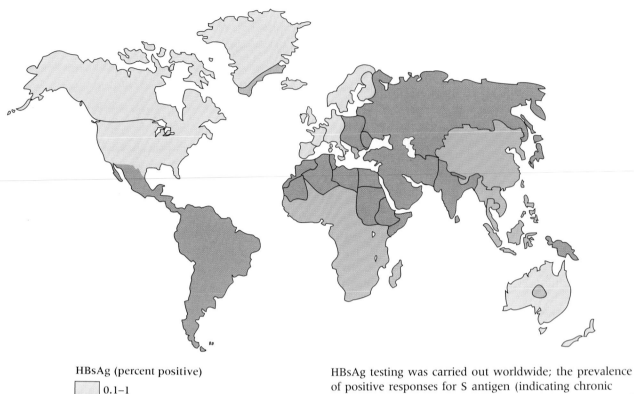

HBsAg (percent positive)

	0.1–1
	1–5
	5–20

HBsAg testing was carried out worldwide; the prevalence of positive responses for S antigen (indicating chronic disease) is shown on this map. Compare the geographical incidence of primary hepatocellular carcinoma (next figure).

The mechanism of liver-tissue damage is unclear. Some evidence indicates that the immune response, with both antibodies and CD8 (killer) T cells attacking viral antigens on the surface of infected cells, kills large numbers of liver cells, reducing the organ's function. The fact that mild or subclinical hepatitis occurs in immunologically impaired patients is consistent with this explanation; in addition, infants mount a less vigorous immune response against infected liver cells than do adults and have less disease. If this hypothesis is correct, it is our own immune response to hepatitis B virus that initiates pathology.

Depending upon a range of variables, most patients manage to overcome the virus infection and restore good liver function. About 2 to 10 percent of adults with hepatitis B infections do not clear the virus or viral proteins; rather, they continue to synthesize virus or viral antigens, in some cases for very long periods of time, and this is termed chronic hepatitis. It is possible to move from either subclinical acute hepatitis (no symptoms) or severe primary-disease symptoms to the chronic phase.

Chronic hepatitis can rapidly lead to cirrhosis (tissue destruction) of the liver and death (about 10 percent of cases) or can remain asymptomatic for years. The chronically infected patients, ill or not, can have high levels of virus in their blood, saliva, semen, or tissues, and represent a large carrier pop-

Annual incidence of primary liver cancer
(cases/100,000 population)

1–3	10–150
3–10	Poorly documented

The annual reported cases of primary hepatocellular carcinoma are indicated on this map. Compare the geographical incidence of hepatitis B virus infection in its chronic form (previous figure).

ulation. Pregnant women can pass the virus to their offspring via the birth canal and, less frequently, in utero. In Asia and some parts of Africa, this may be one of the commonest modes of transmission; for some reason, the rate of infection from mother to offspring is much higher in Oriental populations (40 to 50 percent) than in Caucasian populations (less than 10 percent). Age has a considerable impact upon the development of chronic hepatitis. An individual infected in the first two or three months of life has an 80 to 85 percent chance of becoming a chronic carrier (versus 2 to 10 percent in adult infections). This is thought to result from the different abilities of infants and adults to mount an immune response to the virus and infected liver cells.

The Epidemiology of Hepatitis B Virus

HBsAg testing was carried out worldwide to determine the scope of hepatitis B virus carrier populations. In 1970, when 3.5 billion people inhabited the world, an estimated 176 million hepatitis B virus carriers were detected (5 percent of the population). In China, Taiwan, Southeast Asia, sub-Saharan Africa, Alaska (Inuit), and along the coast of Greenland, between 5 and 20 percent of the people had persistent hepatitis B virus infections. In the Middle East, North Africa, South and Central America, southern and eastern Europe, and the

USSR, between 1 and 5 percent were hepatitis B virus carriers. The lowest rates—0.1 to 1.0 percent —of this virus infection were localized in North America, Scandinavia, Australia, and New Zealand.

These high transmission and persistence rates in Asia follow a vicious circle of family infection and reinfection. Husbands (5 to 20 percent of this group) transmit the virus to their wives in semen, and the pregnant wife will pass the virus to her children with very high efficiency. Childhood infection ensures that a high percentage of the adults in the population will have the chronic, persistent form of virus production and disease. This cycle explains some of the geographical distinctions in incidence and prevalence of the disease. Often local traditions, such as the chewing of food for an infant by the mother (the virus is in saliva), may enhance the spread of this virus.

Besides these natural routes of virus spread, hepatitis B virus was a common contaminant of the blood supply until tests for the virus, viral carriers, and HBsAg virtually eliminated it from this source. Drug addicts who share needles, however, still spread hepatitis B virus, and venereal transmission has led to high rates of hepatitis B virus and chronic infection in prostitutes and in young homosexual men with multiple partners. About 7 percent of the homosexual men in the United States are HBsAg-positive. As mentioned earlier, hepatitis B virus is endemic in residential facilities for the mentally handicapped, probably reflecting crowded conditions and, in some cases, poor hygiene.

The Association of Hepatitis B Virus with Hepatocellular Carcinoma

The incidence of primary hepatocellular carcinoma in the world closely follows the incidence and persistence of hepatitis B virus. A number of studies have demonstrated that a chronic carrier of hepatitis B virus has a much greater than normal risk of developing liver cancer. In a detailed prospective study carried out by R. P. Beasley and his colleagues in Taiwan, 22,707 Chinese men were followed over six years. Of this group, 3454 were HBsAg-positive (15.2 percent). In the six-year period, 116 died of hepatocellular cancer, and all but three of these patients were in the HBsAg-positive group. The risk of developing liver cancer was 105 to 217 times greater for HBsAg carriers than for noncarriers. On the average, one in 322 HBsAg carriers died of liver cancer per year. From this study, it could be predicted that the likelihood of a Chinese HBsAg-positive male carrier developing primary liver cancer during a fifty-year period is about 15 percent. In the group of HBsAg carriers, the annual incidence of hepatocellular cancer was constant over five years, indicating that new liver cancers were arising each year in the HBsAg-positive population. Deaths from liver cancer in this group increased as a function of their age. A very similar relationship was demonstrated between the numbers of HBsAg carriers and those who died from cirrhosis of the liver.

Thus it is clear that chronic hepatitis is a very large risk factor for liver cancer. The question that remains is why. Some scientists have pointed out that a patient with chronic hepatitis consistently experiences episodes of virus and HBsAg synthesis, followed by killer T cells attacking the infected liver cells. This results in liver damage, regeneration of lost cells, and reinfection of the new cells with virus, setting up a cycle of constant liver-cell regeneration and division over a lifetime of persistent infection. Dividing cells can make mistakes during DNA replication, so there is a higher mutation rate in dividing cell populations than in resting ones. Mutations located in potential oncogenes would contribute to cells that become cancerous. The cycle of cell death and regeneration, therefore, puts the host at risk for a higher incidence of cancer. The role of the virus is to promote this cycle.

Alternatively, other scientists believe that the X gene product, which may promote transcription of some viral and cellular genes, might be responsible for altering cellular gene expression, leading to cancer. The advocates of this idea point out that a number of other viruses encode proteins that play a role in transforming cells in culture or initiating tumors in animals (the E1a genes of adenoviruses and the tat gene of HTLV-I are the prototypes). While this is an interesting hypothesis, there is no evidence to support it. All of the DNA tumor viruses, which were discussed in Chapter 5, possess a set of genes that are properly termed viral-encoded oncogenes: they produce proteins that directly alter the growth or replication of cells. That does not appear to be the case for hepatitis B virus (the X gene does not transform cells in culture). The mechanism by which this virus, in a chronic carrier, predisposes the host to cancer remains one of the mysteries for the next generation of virologists and cancer biologists to solve.

Hepatitis B Virus Vaccines

As is evident from the previous discussion, the hepatitis B virus has an enormous impact upon the peoples of Asia and parts of Africa. Worldwide, there are an estimated five hundred thousand deaths annually from liver cancer, most of which are hepatitis B related. In the industrialized Western nations, liver cancer represents only 2 to 3 percent of total gastrointestinal cancers. In the United States there are five thousand cases of liver cancer per year in a population of 250 million, while in Taiwan there are ten thousand deaths a year from hepatocellular carcinoma in a population of seventeen million. Liver cancer is the most frequent cancer in males in many parts of Southeast Asia and Africa. How can we break the cycle of virus infection at birth leading to death by liver cancer forty or fifty years later? The answer may well have come in the form of a hepatitis B virus vaccine.

The approach to making such a vaccine was to obtain HBsAg in pure form, separated from the infectious Dane particles; because HBsAg induces neutralizing antibodies for the virus, it would be a very suitable vaccine. The largest source of HBsAg was the serum of chronic persistent carriers, which could yield up to one trillion to ten trillion HBsAg particles (22-nm particles of membrane and S antigen) per milliliter. Scientists at the Merck, Sharp, and Dohme pharmaceutical company succeeded in purifying these HBsAg particles from infectious hepatitis B virus and other possible contaminating viruses in the blood of patients. This is the only plasma-derived hepatitis B vaccine manufactured and licensed in the United States at this time; internationally, there are about twelve manufacturers of the plasma-derived vaccine. More than thirty million doses of this vaccine were administered throughout the world between 1978 and 1988, and the safety record is excellent.

In November 1978, one of the first trials to test the efficacy of this vaccine was begun. HBsAg was given to 549 male homosexuals who were HBsAg-antibody-negative; they had no prior exposure to the virus but were in a high-risk group. As a control, 534 homosexual men received a placebo. Three shots are required, the second after one month and the third at six months. Individuals receiving HBsAg made antibodies to this antigen: after the first injection, 31.4 percent made antibodies; after the second, 77 percent; and by nine months (three months after the third injection), 96 to 98 percent. Hepatitis B virus infected 25.6 percent of the placebo group and 3.2 percent of the vaccine recipients. All those subjects who had vaccine exposures and subsequent hepatitis B virus infections (the 3.2 percent) were in a group that failed to make antibodies to HBsAg. Subsequent clinical trials also confirmed the utility of this vaccine.

A second-generation hepatitis B virus vaccine has recently been developed. The goal was to find a source of HBsAg other than human plasma from hepatitis B virus carriers—it was expensive and somewhat hazardous to handle large volumes of contaminated human plasma, and there remained the possibility that infectious virus might contaminate some lots of vaccine. In the 1980s, with the advent of HIV in some of the same individuals who were HBsAg-positive, it seemed wise to develop an alternative approach. A recombinant DNA vac-

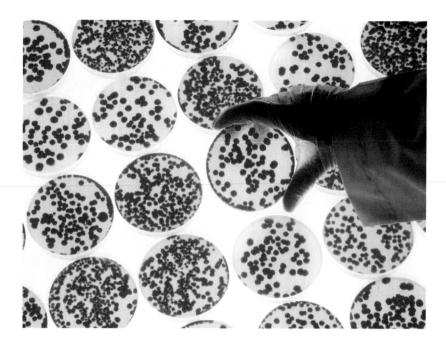

Modern pharmaceutical laboratories employ a hundred years' worth of biotechnologies—from growing organisms in petri dishes (shown here) to the recombinant DNA approach that recently yielded a safe HBsAg immunizing agent.

cine was produced by Chiron together with Merck, Sharp, and Dohme. The gene for HBsAg protein was isolated and expressed in yeast, which could be grown in large fermenters. The yeast synthesized HBsAg and inserted it into a membrane vesicle that could be isolated and purified. The 20-nm antigenic particles were almost identical to the HBsAg from plasma, and they immunized individuals as efficiently.

Although this vaccine can protect individuals before exposure to hepatitis B virus, the problem of the exposure of newborns and the cycle of virus infection in Asia remained. Could the vaccine protect infants exposed to their mother's virus—and, if so, would this break the inexorable round of infection, lifelong carrier state, and liver cancer? In July 1984, the Taiwanese government launched a mass immunization program. Over fifteen months, 352,721 pregnant women (about 78 percent of all pregnant women in Taiwan during that period) were screened for HBsAg in their blood. Eighteen percent—62,359—were HBsAg-positive, and at least half of these women were highly infectious,

producing large levels of Dane particles. The infants born to the HBsAg-positive women were given the hepatitis B virus vaccine at one, five, and nine weeks after birth, with a booster shot at twelve months. All the infants whose mothers had high levels of infectious virus also received a shot of immunoglobulin (antibody directed against HBsAg, to neutralize the virus if possible) within twenty-four hours after birth.

When these children were about eighteen months old, 3464 randomly selected infants in the study were tested for the presence of (1) HBsAg, which would indicate continued virus infection; (2) HBsAg antibodies, which would indicate a good response to the vaccine and protection against the virus; and (3) HBcAg antibodies, which would indicate an immune response to a hepatitis B virus infection (HBcAg is not present in the vaccine, but patients exposed to the virus make this antibody). Of 786 infants whose mothers had high levels of infectious virus and who therefore received both immunoglobulin and the vaccine, about 85 percent were now virus-free and protected. Overall, only

11 percent of the infants still carried HBsAg in their blood. If these are representative numbers, the immunization program should reduce the hepatitis B virus carrier rate in the next generation from 18 percent to about 2 percent; forty to fifty years from now, 8300 fewer hepatocellular carcinomas will be reported than would have been expected—and that will be observed in people born in 1983, the year before the immunization program began. The first human vaccine designed to reduce the incidence of liver disease and cancer is off to a good start.

Slave deck of the Albaroz, Prize to the H.M.S. Albatross.

The Origins and Evolution of Viruses

Nature is a very strange affair, and the strangenesses already encountered by our friends the physicists are banalities compared to the queer things being glimpsed in biology, and the much queerer things that lie ahead.

As these turn up . . . they will inevitably change the way the world looks. And when this happens, the view of life itself will also shift; old ideas will be set aside; the look of a tree will be a different look; the connectedness of all the parts of nature will become a reality for everyone, not just the mystics, to think about; painters will begin to paint differently; music will change from what it is to something new and unguessed at; poets will write stranger poems; and the culture will begin a new cycle of change.

Lewis Thomas, 1985

Left: Ships carrying slaves to the Western Hemisphere also imported yellow fever, whose agent—a highly adaptive virus with both primate and mosquito hosts—was the first viral isolate of human disease identified. British Lt. Francis Meynell (1821–1870), who spent the years 1844 to 1850 aboard the HMS *Penelope* hunting slave traders along the Atlantic coast of Africa, kept a sketchbook. Here he depicts the slave deck of the *Albanez,* captured by HMS *Albatross.*

Above: EM of eastern equine encephalitis virus. Like its relative, the yellow fever virus, this virion has a ribonucleoprotein core surrounded by an outer envelope with glycoprotein spikes, acquired by budding from the plasma membrane of an infected cell (bottom of EM). These viruses replicate in both mammals and insects.

All life forms are composed of two classes of chemical elements: those employed in information storage and those used for functioning. In all organisms, information is stored in the nucleotide sequences of nucleic acids: DNA for most living forms, RNA for some viruses. The functions encoded by these sequences are carried out by and large by proteins, whose properties are determined by a genetic code common to all life on Earth. As very small obligate intracellular parasites with limited capacity for information storage, viruses are among the most efficient and economical forms of life. Viruses encode the information only for those functions that they cannot exact

195

from their hosts, so they waste few nucleotides in their genomes. Some viruses (hepadnaviruses, retroviruses, and SV40, for example) even use the same nucleotides to encode more than one protein, using alternative reading frames (see page 184), while other viruses employ mRNA splicing to maximize the number of different proteins that they can synthesize.

We have seen that viruses' gene functions fall into two categories—those employed for replication or to carry out a particular stage in the life cycle (latency, lysogeny); and those used to package or assemble the genetic information into virus particles. Inherent in the structure of a virus particle are two competing requirements: a virus shell must be sufficiently stable to resist the environmental insults it is exposed to outside a cell—temperature, solvents, drying; but it must also be capable, upon entry into a cell, of rapid disassembly and an orderly readout of the information stored in its genome. Despite limits on their information storage, size, structure, and so forth, viruses have clearly solved these problems and, indeed, have often thrived.

We have come to the place where we can productively ask several questions: How did viruses originate? What are the forces that create new viruses and the selection pressures that result in new diseases? How do viruses change and evolve? But we must appreciate our limitations. Viruses have left no fossil record. Even more problematic is the fact that the oldest viruses we can examine were only discovered eighty to a hundred years ago. This is not an evolutionary time scale, even with a short generation time and a high mutation rate. So we must use present-day viruses and compare related families and distant cousins. From a detailed analysis of today's viruses, we may extrapolate back and project forward to possible patterns of change and evolution. Although it is important to realize that we cannot expect definitive answers to our questions, we can hope to provide a sound basis for making future decisions about viruses and the diseases they cause.

Theories to Explain the Origin of Viruses

In addressing the origin of viruses, we face some of the same problems tackled by Robert Koch in the late nineteenth century. He had to provide a set of rules by which to recognize whether an organism causes a disease. In the context of this discussion, we need to be able to recognize an origin event and to be clear about what we mean by *originate*. Let us define the origin of a new virus as that time when its replication and evolution become independent of the molecules from which it was derived. When a genetic element acquires the information needed to duplicate itself and determine its own destiny, it has achieved the status of a new and independent life form. With this definition in mind, we can now search for origin events. The places we should be looking vary with the theories of how viruses arise.

Three different theories have been proposed to explain the origins of viruses. The first, the regressive theory of virus origins, proposes that viruses arise from free-living organisms like bacteria that have progressively lost genetic information—to the point where they become intracellular parasites dependent upon their hosts to supply the functions they have lost. The second theory proposes that viruses arise from the host-cell RNA or DNA molecules, which gain a self-replicative but parasitic existence. One or a few genes—or the mRNA from one or a few genes—acquires the ability to replicate and evolve (change its nucleotide sequences or organization) independently of its host gene or mRNA. According to this hypothesis, viruses arise directly from the host cell. By contrast, the third theory proposes that viruses originated and evolved along with the most primitive molecules that first contained self-replicating abilities. While some of these molecules were eventually collected into units of organization and duplication termed cells, other molecules were packaged into virus particles that coevolved with cells and parasitized them. Let us review each of these theories in greater detail.

Schematic representation of three theories:
How did viruses evolve?

POSSIBLE ORIGINS OF VIRUSES

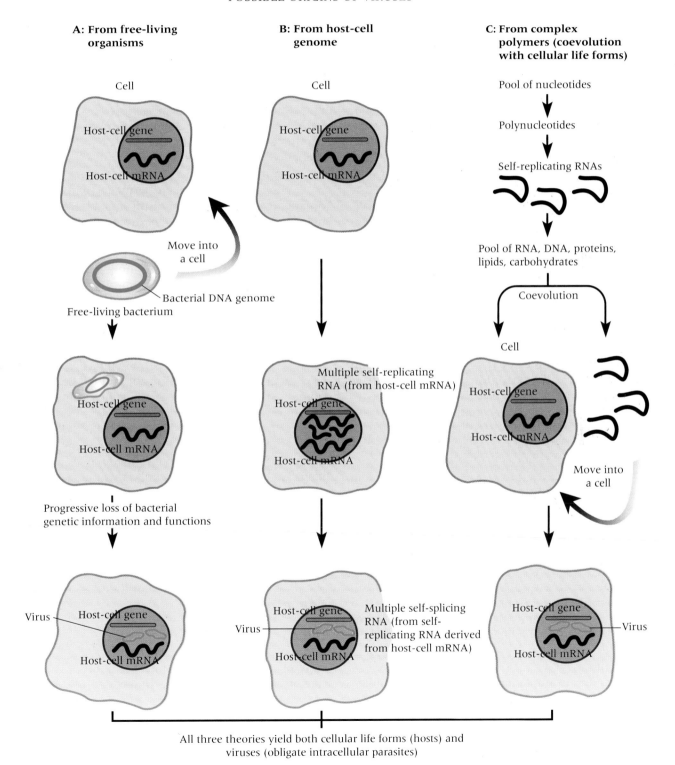

A: From free-living organisms

Cell

Host-cell gene

Host-cell mRNA

Move into a cell

Bacterial DNA genome

Free-living bacterium

Host-cell gene

Host-cell mRNA

Progressive loss of bacterial genetic information and functions

Virus — Host-cell gene

Host-cell mRNA

B: From host-cell genome

Cell

Host-cell gene

Host-cell mRNA

Multiple self-replicating RNA (from host-cell mRNA)

Host-cell gene

Host-cell mRNA

Virus — Host-cell gene

Multiple self-splicing RNA (from self-replicating RNA derived from host-cell mRNA)

Host-cell mRNA

C: From complex polymers (coevolution with cellular life forms)

Pool of nucleotides

Polynucleotides

Self-replicating RNAs

Pool of RNA, DNA, proteins, lipids, carbohydrates

Coevolution

Cell

Host-cell gene

Host-cell mRNA

Move into a cell

Host-cell gene

Host-cell mRNA — Virus

All three theories yield both cellular life forms (hosts) and viruses (obligate intracellular parasites)

Do Viruses Originate by Regression from Free-living Organisms?

Many free-living bacteria use glucose, simple salts, and water as energy and as building blocks for replication. These bacteria are single cells that multiply by duplicating their components and splitting in half. Some microorganisms are more fastidious, requiring preformed metabolites—vitamins, amino acids, and so on—which they get from their environment. Intracellular parasitism, or the movement of an organism into a cell to obtain these preformed metabolites, removes a number of constraints from such an organism. Once intracellular life is established, the loss of further genetic information and biosynthetic systems is no longer disadvantageous; the host cell can supply most of the needs of a bacterium that has taken up life inside the cell. The only essential functions that this parasite must maintain are (1) an ability to replicate itself and regulate that replication, and (2) a way to interact with its host and regulate its own needs. The progressive loss of information (DNA) and the evolutionary changes made to adapt to this new life style create a new organism.

Several examples of present-day intracellular parasites are clearly derived from free-living bacteria. The rickettsiae, which cause such diseases as typhus, are small bacteria that are obligate parasites within cells. They multiply by fission (like bacteria and unlike viruses) and have lost many of their biosynthetic enzymes and capabilities. Perhaps more like viruses are the chlamydiae, organisms that cause inflammations of the eye and the urinogenital tract in humans; they have no cell walls (bacteria have such protective walls) and cannot live outside of cells. Chlamydiae are like bacteria in that they contain both DNA and RNA—viruses have only one or the other—divide by fission, and have some cellular structure. For a long time this group was classified as viral, but it is now quite certain that they are the remnants of a degenerative process starting from larger, more complex bacteria.

This continuing process of information loss may well have given rise to the eukaryotic cell organelles: for example, mitochondria and chloroplasts. These organelles are about the same size as bacteria, contain circular DNA genomes like bacteria, and encode the information to synthesize ribosomal RNAs and transfer RNAs. Like bacteria, these organelles contain both DNA and RNA; they even have an independent way to synthesize proteins encoded by the organelle DNA. A quite noticeable nucleotide-sequence similarity exists between the ribosomal RNA in bacteria and in plant mitochondria. Such nucleotide resemblances are consistent with an evolutionary relationship, but we cannot eliminate the possibility that parallel nucleotide sequences evolved independently in each instance to do parallel jobs.

While a good case can be made that the cell organelles of higher organisms may have originated from bacteria by degeneration, it is less clear that viruses arose in this way. Viruses never encode ribosomal RNAs; nor do they bring to a cell their own machinery to synthesize proteins, although some viruses do have genes for transfer RNAs. Further, we know of no living intermediates between chlamydiae and viruses. The eukaryotic cell organelles, in common with the bacteria, have only DNA genomes, so it remains unclear from this theory how RNA viruses might arise. Finally, the regressive theory may explain how a simple DNA genome and an intracellular parasitic life could begin, but it does little to suggest how the next step—the formation of a coat or shell about this DNA genome—might be accomplished. The genetic information in a cell might be employed as a starting material to package the viral DNA, but exactly what might drive this process and what its selective advantages could be remain less clear. The degenerative theory of virus origins appears to leave a number of questions unanswered.

Do Viruses Originate from Components of Cellular RNA or DNA?

This theory postulates that a portion of the genetic information found in the host cell, or an mRNA copy of this DNA, can acquire the ability to replicate itself and then evolve independently of the original host-cell genome. This constitutes an origin event. The question then becomes: What is needed to create a self-replicative molecule that is free to evolve independently over time? All living organisms must duplicate the molecules that store their information (DNA or RNA). They do so by an ordered process that starts at a unique site on the polymers and ends when all the nucleotides are duplicated in the correct sequence. The start site position has been termed an origin of replication, and every self-duplicating molecule must have one. The nucleotide structure at the origin is recognized by one or more proteins whose function is to initiate polymerization of the nucleotides into a complementary molecule. By obtaining a new nucleotide sequence recognized as an origin of replication, an RNA or DNA molecule synthesized by a host cell may acquire the ability to duplicate itself independently.

If we look for molecules with these properties, we can find them easily in bacterial cells; they are called episomes. Episomes are circular DNA molecules that can exist in an autonomous state—replicating freely in the cytoplasm, where they are termed plasmids—or in an integrated state contiguous with the bacterial chromosome. In this sense they are similar to the lambda bacteriophage, which exists in both states; its DNA is also termed an episome. Whether the episome is free or integrated into the DNA of its host, its nucleotide sequence duplicates once per host-cell generation and is segregated into the two daughter cells.

The Continuum of Molecular Life Forms

Bacterial chromosome	3×10^6 nucleotides; contains the information for the great majority of bacterial functions (about 5000 genes)
Episome	Circular DNA molecule, variable in size (nucleotide number); present in an autonomous (plasmid) or integrated form; can carry the genes for antibiotic resistance, fertility, and so forth
Transposon	DNA molecule, variable in size (750–40,000 nucleotides); can move from one integrated location in a plasmid or bacterial chromosome to another; can carry with it other genes (for example, for antibiotic resistance); composed of terminal repeats that are insertion elements
Insertion (IS) element	DNA molecule, variable in size; these nucleotide sequences encode information for moving the element from one chromosome location to another
Retrotransposon	An IS element that requires DNA-to-RNA-to-DNA steps for movement of the element; encodes its own reverse transcriptase
Retrovirus	The provirus (DNA) has terminal repeats and encodes a reverse transcriptase; the gag and env gene functions permit the virus to move from cell to cell (that is, to be infectious)
Hepadnavirus	DNA virus that replicates by means of a DNA-to-RNA-to-DNA step, using a polyermase-reverse transcriptase enzyme
Lysogenic phage	Can exist in an autonomous or an integrated state
Plasmid-virus	The lysogenic phage P2 can package the plasmid P4, converting it to a virus with P2's own coat proteins

Plasmids can be quite useful to their host cells. Moving from one bacterium to another by a process of conjugation (genetic cross), episomes can continue to acquire new genes and functions. One common set of genes carried by plasmids results in the bacteria becoming resistant to antibiotics; these genes make proteins that degrade or alter the antibiotic and destroy its activity. Indeed, the genetic information that permits a bacterium to act as a male and conjugate by donating its chromosome is encoded by a set of genes that reside on episomes. The fact that plasmids have their own genes (for functions like antibiotic resistance and fertility) not found on the host chromosome means that they evolve independently of their hosts.

One way plasmids evolve is to acquire entirely new sets of genes, which are often inserted into plasmid chromosomes by another kind of genetic element, a movable one called a transposon. Transposons are DNA elements of 750 to 40,000 nucleotides that have the ability to move from one chromosomal location to a new site in another chromosome. In so doing, transposons can carry with them banks of genes for antibiotic resistance and other functions.

All transposons have a common structure. The bank of genes to be moved is bounded on both sides by a DNA sequence termed an insertion (IS) element. IS elements are DNA sequences that encode the information for their own movement from one place on a chromosome to another. We now recognize three basic classes of IS elements: (1) one that cuts itself out of the DNA chromosome and moves to another site, where it inserts; (2) one that duplicates itself and uses the duplicated DNA copy to move to another location on a chromosome; and (3) one that makes an RNA copy of its own DNA, which in turn is copied into DNA by reverse transcriptase, followed by a new insertion of this DNA copy into a chromosome. This third class of IS elements, commonly called retrotransposons, encodes a gene for the reverse transcriptase enzyme. Very closely related to the retroviruses and hepadnaviruses, retrotransposons have been found in yeast, flies, and higher organisms, so this is not an uncommon phenomenon.

What becomes clear from this discussion is that there is a series of DNA and RNA molecules—episomes and IS elements—that have separate replicating systems, evolve independently of their hosts, and yet live in either an integrated or an autonomous form: sometimes part of the host cell and sometimes free. Even more striking, some episomes are also the genomes of viruses (lambda phage), and some transposons are closely related to viruses (hepadnaviruses, retroviruses). In the continuum of these molecular life forms, episomes, plasmids, IS elements, and viruses seem to arise from each

This EM shows the long, linear, double-stranded DNA of bacteriophage T7 (39,936 nucleotides and 14 microns long) as a size standard. T7 DNA encodes the information for 55 proteins. The potato spindle-tuber viroid shown is a single strand of circular RNA that—by extensive self-complementary base pairs—forms a double helix about 359 nucleotides (0.05 micron) in length. This organism contains no genes and does not encode the information for any protein; it is simply a set of nucleotide signals capable of self-replication with the help of its host (a potato).

The nucleotide sequence of potato spindle-tuber viroid—first determined by Dr. H. J. Gross and his colleagues at the Max Planck Institute in Munich, Germany—is presented and folded into a maximally base-paired structure. Its single-stranded, circular RNA is composed of 73 A, 77 U, 101 G, and 108 C nucleotides (note the excess of G and C and the RNA self-replicating molecules), extensively base-paired.

other; the distinctions between these molecules rapidly disappear as life styles evolve.

Yet another set of origin events can be imagined, based upon some of the unique properties of RNA. While we have been distinguishing between molecules that store information (DNA, RNA) and molecules like proteins that function to promote chemical reactions, RNA is unique in that it can both store information in its nucleotide sequence and, under some circumstances, act as a biochemical catalyst: an enzyme. We know of three chemical reactions that can be promoted by RNA. (1) Some RNA molecules can cut a long polynucleotide chain at a specific site. (2) Some RNA molecules can splice out an internal sequence of nucleotides in a polymer, deleting them. This so-called self-splicing requires cutting and reforming the continuous chain of the polymer. (3) Some RNA molecules can promote a template-dependent synthesis of new polymers. For example, a polymer containing many nucleotides of cytosine (CCCCCCCC) can synthesize a polymer of guanine (GGGGGGGG) with which it pairs, making a double-stranded RNA helix. This remarkable capacity to generate a complementary nucleotide sequence means that RNA both stores the sequential information and actively participates in its new duplication.

These three functions—duplication; cutting at specific sites; splicing and patching—if carried out by RNA molecules in isolation, may be enough to constitute an origin event. Indeed, it is hypothesized that the very first forms of life began with RNA polymers, in a primordial "soup"; simultaneously informational and functional, they duplicated themselves. The first living world, according to this view, was an RNA-based world; but DNA eventually replaced RNA as the information molecule because a nonreactive, stable molecule was more desirable for storage. The very reactivity of RNA threatens its permanence: we have seen that RNA viruses evolve more rapidly than do DNA viruses.

If these three chemical reactions, carried out by RNA, indeed played a role in the origins of life, can we find evidence today of life forms with the expected properties of these molecules? Such organisms do exist; called viroids and virusoids, they are the smallest known autonomously replicating molecules. Viroids are plant pathogens composed of a single-stranded RNA molecule 240 to 375 nucleotides in length. The nucleotide sequences in this circular RNA contain many regions of self-complementarity (that is, G = C and A = U pairs are formed), and the molecule is extensively base-paired over its entire length. These nucleotide sequences do not appear to encode any protein, so replicative functions are supplied entirely by the host cell. Viroids have only one unique origin of replication (start site), plus the ability to carry out self-splicing reactions to produce a molecule of a defined length from a longer, newly duplicated precursor. Inasmuch as the size and structure of the viroid RNA are controlled by its nucleotide sequences, the viroid evolves independently of its host cell and is alive—it has had an origin event.

Viroids replicate in plant-cell nuclei and often cause striking abnormalities or diseases in their host plants. They have no coat proteins, and the

Camellia leaf showing distinctive pattern of plant damage that results from viroid infection, which is transmitted in the course of horticultural grafting.

RNA does not readily leave one cell or plant and infect another. Because of this, one might imagine viroids to be at a disadvantage in spreading to new hosts. In relatively modern times, however, humanity has come to their rescue. By planting fields of a single species, by grafting one plant part onto another plant (the knife carries the RNA), and by other agricultural practices, we have spread viroids from plant to plant, seed to seed, and place to place—in some cases causing economic devastation. Palm trees, tomatoes, and potatoes are examples of crops that have proved vulnerable to serious damage from viroids.

Virusoids are the viruses of viruses—parasites of a helper virus. Virusoids, like viroids, are small (200 to 400 nucleotides), circular RNA molecules that are extensively base-paired into a double helix. They are always found with a larger RNA plant virus that encodes the genetic information to replicate this smaller RNA and package it into virus particles. Virusoids duplicate themselves in the cytoplasm and undergo a self-splicing reaction that

determines their size. Their origin of replication, a nucleotide sequence, must be recognized as a start site of duplication by an RNA polymerase encoded by the helper virus.

The structure of virusoids and viroids—a single-stranded but base-paired, circular molecule of a unique size—is just what one might predict as the product of an RNA template-dependent synthesis and self-splicing reaction. If one examines the nucleotide sequences of several types of viroids, it becomes clear that regions of their chromosomes retain a common, closely related nucleotide sequence, while the nucleotide sequences of other regions differ widely among different viroid molecules. This may indicate both the conservation of functionally useful nucleotide sequences (origin of replication; self-splicing functions) and an evolutionary process of viroid change as the organisms find new hosts and adapt to new environments. If this is correct, then origin events for new viruses could be occurring all the time. The existing reservoir of hosts, cells, genes, and molecules would constantly be used to generate origin events and new forms of life. In this sense, the creation continues to this day.

Do Viruses Originate from Self-Replicating Molecules?

As we have seen, it appears that the earliest self-replicating information systems were composed of RNA polymers. While the chemical reactions for duplicating and cutting the RNA polymers are promoted by the RNA molecules themselves, the rate at which this happens is slow—only about two monomeric nucleotides are added to a growing chain per minute at 25°C. This contrasts with the rate of nucleotide addition into RNA copied from a DNA template by the enzyme RNA polymerase—fifty nucleotides per second at the same temperature. The protein is a more efficient catalyst: 1500 times faster.

Proteins were also among the chemical components of the early world when life began and

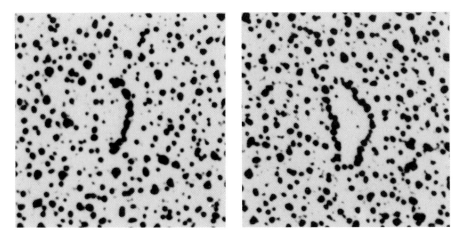

EMs of a potato spindle-tuber viroid molecule magnified some 440,000 times. This viroid, with its 240 base pairs (equivalent to computer bits), contains about ten million times less genetic information than the human genome. Right: Circular viroid RNA under conditions that dissolve hydrogen bonds, preventing A = U and G = C base-pairing. Left: The same molecule when it is fully base-paired (hydrogen-bonded), looking like a double-stranded rod.

evolved. This theory proposes that because proteins are more efficient in promoting important chemical reactions useful to life forms, the initial RNA information-duplication systems evolved into RNA-protein information-replication processes. Probably for the sake of better stability and reproducibility, DNA came to replace RNA as the information storage molecule, and RNA assumed the role of intermediate or messenger, while proteins carried out the functions. Once the sequence of nucleotides in DNA determined the sequence of amino acids in proteins, a genetic code common to all life forms on Earth was fixed, becoming the language used by this two-part system of life.

As the series of living molecules became more complex, it was advantageous to package the genetic information and its readout systems. Lipids, insoluble in the aqueous cell's environment, surrounded these molecules, separating them from the outside: in this way, some life forms became cells. Other life forms may well have been simple nucleic acids surrounded by protein coats. To duplicate themselves, these molecules entered a cell. In this way, host cells and viruses might both have arisen in the same primordial "soup." It is reasonable, according to this view, to suppose that such self-replicating molecules—the viruses—and their host cells coevolved. While new origin events may still occur to generate viruses, spontaneous origin events that result in more complex organisms like bacteria no longer occur—as shown by Pasteur's famous nineteenth-century experiments refuting the theory of spontaneous generation.

Factors Affecting the Evolution of Viruses

While it may not be possible to provide definitive answers to the question of how viruses originated, the forces that drive the evolution of viruses are much clearer. Two major mechanisms act upon the genetic information of a virus (or any life form) to produce change: mutation and recombination. Mutation is a change in the sequence of the nucleotides in DNA (A, T, G, and C) or RNA (A, U, G, and C) polymers. Recombination is the exchange and bringing together of new sequences of nucleotides, in new combinations, from two parental polynucleotide strands of DNA or RNA. Both processes generate diversity in all living organisms, which can then be tested for environmental advantages and replicative fitness in the real world.

Because viruses can encode the enzymes for their own replication and recombination, some viruses have a good deal of control over their mutation rates and frequencies of recombination. In these cases, the properties of the viral protein itself (DNA or RNA polymerase) dictate (1) error rates in polymerization, (2) potential to recognize and correct errors, and (3) recombination rates—all of which vary among different viruses. Mutation rates of RNA viruses are higher than those of DNA viruses, because the polymerase enzymes that duplicate RNA into RNA—or even RNA into DNA (reverse transcriptase)—make mistakes more often than do DNA polymerases or fail to repair mistakes when they are detected. Some viruses encode a set of genes for proteins that promote recombination in an infected cell. If there are a hundred copies of a viral DNA chromosome in a cell, and each copy is slightly different because of mutation errors during replication, an active recombination system will allow the creation of a single chromosome containing two or three of the changes previously present on separate DNA polymers. New combinations of changes arise and are tested for their suitability. It will be instructive to take a look at some examples of how recombination and mutation act to create genes, viruses, and even new diseases.

The Acquisition of Foreign Genes

In earlier chapters we have made a sharp distinction between the prokaryotic world of bacteria, which lack membrane-enclosed nuclei, and the eukaryotes. In prokaryotes, each gene is composed of a simple, linear readout (array) of nucleotides that determines the linear order of amino acids in a protein. In many eukaryotic genes, however, the coding information for a given protein is interrupted by nucleotide sequences that encode no information for that protein—a eukaryotic gene, then, may not contain a simple, linear readout of information. Rather, the nucleotide sequence of the gene is first copied into RNA; then the RNA undergoes a series of splicing reactions that remove those nucleotides not encoding information used to make a protein. The coding regions of a gene are called exons, and the noncoding regions are termed introns; it is the introns that have been removed by splicing in the final mRNA, so that the messenger RNA of a eukaryotic cell looks like the mRNA of a prokaryotic (bacterial) cell.

Because the bacteriophages live and replicate in a world of prokaryotic host cells, their genes are also simple, linear readouts of nucleotide sequences. It was a big surprise, therefore, when it was found that the bacteriophage T4 (see Chapter 2) contained several genes that include internal nucleotide sequences that must be spliced out of the RNA before it can function properly. The host bacterium, *E. coli*, does not contain any genes or functions to aid in gene splicing, because no other genes like this exist in prokaryotes. The phage RNA—like the viroids and virusoids of plant cells— can undergo self-splicing (either spontaneously or when mediated by a protein encoded by the intron). Genes like this are commonly found in eukaryotes, but never in these host bacteria. Where did the T4 genes come from? How did eukaryotic genes or intron elements get into a prokaryotic world?

There are two possibilities. First, they perhaps evolved in the T4 chromosome by a series of mutations that just happened to produce a self-splicing RNA. An insertion element could produce such a split gene. Once produced, it could have been selected for; it was useful. Duplication and transposition of this element would then make the several copies of it that are located in different places on the T4 chromosome. Alternatively, the gene that

PROKARYOTE   EUKARYOTE

DNA

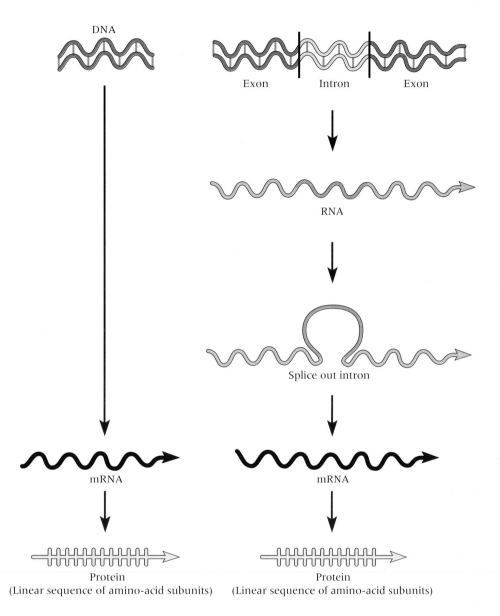

Exon   Intron   Exon

RNA

Splice out intron

mRNA   mRNA

Protein   Protein
(Linear sequence of amino-acid subunits)   (Linear sequence of amino-acid subunits)

Three nucleotides in DNA or mRNA   Three nucleotides in exon DNA and mRNA
determine one amino acid in a   determine one amino acid in a protein
protein polymer.   polymer (the intron mRNA is spliced out).

Schematic representation of prokaryotic and eukaryotic genes. The interrupted
gene or split-gene structure—exon-intron-exon—is absent in prokaryotes.

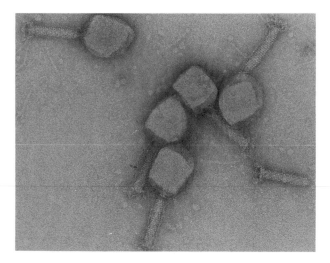

EM of bacteriophage T4. All the genes in T4's bacterial host, *E. coli*, are uninterrupted linear sequences of nucleotides that encode the information for proteins; they do not contain introns and have no known mechanisms to splice mRNA. But three genes in T4 contain internal sequences (introns) that must be spliced out of the RNA to synthesize a product. Where did these eukaryotic-like genes come from?

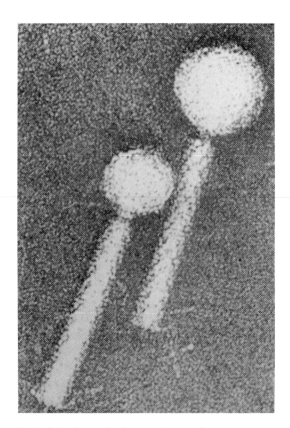

EM of the bacteriophages P2-P4. The P4 episome has been converted—via recombination or mutation—to a helper-dependent virus, with a smaller head and the same tail. Even viruses (P2) have viruses (P4).

creates an intron might have been acquired from a eukaryote. *E. coli,* the host of T4, lives in the intestine of human beings, cows, and many other vertebrates. It is possible that a T4 virus replicated in *E. coli* that had taken up some eukaryotic DNA; recombination could insert such a human or bovine gene into the virus. Recombination permits a living world of genes and elements to be incorporated and tested in a number of different viruses; in this way, it is possible for humans to contribute to the evolution of a bacteriophage.

The Creation of a New Virus

Our second example demonstrates the close relationship between episomes and viruses and shows just how simple it is to convert the former into the latter. Electron micrographs of an *E. coli* bacterio-

phage called P2-P4 show two kinds of virus particles, each with a head and a tail; the tails of both virus particles are identical (135 nm long), but the head of P2 is larger (62 nm in diameter) than that of P4 (45 nm in diameter). These viruses, commonly observed together, have very different capabilities. If P4 enters an *E. coli* host cell, its DNA can replicate and even integrate at a specific spot in the host-cell chromosome. No virus particles are produced, because P4 has only the genetic information to replicate itself or integrate; in other words, it acts like an episome or plasmid of *E. coli*. P2, on the other hand, has a set of gene products to replicate its DNA as well as a set of genes to produce all the structural proteins for the heads and tails of its viral particles; it is a full-fledged virus.

At some time in the past, the P4 DNA acquired (via recombination) a specific nucleotide sequence from a P2 virus that allowed the P2-virus coat proteins in an infected cell to attach to and package a P4 episome. Viruses use such a sequence—often called the packaging nucleotide sequence—to ensure that their own chromosome is placed in a virus particle without wasting valuable coat-protein subunits on foreign DNA. P4, having captured the packaging sequence of P2, was converted from an episome into a helper-dependent phage in P2-infected cells.

P4 is now a virus of P2, which is in turn a virus of *E. coli.* P4 DNA contains 10,000 nucleotide pairs, while P2 DNA has 30,000. The chromosomal volume of a P4 phage head is only one-third that of the P2 head, so two morphologically distinct heads are used to package these two different genomes. Having begun life as a coatless episome, transmitted mostly by conjugation, P4 has found a way to enter a new host, in a "stolen" virus coat. It has changed its life style simply by acquiring a few nucleotides amounting to a packaging signal.

The Final Problem for d'Hérelle: Bacteriophage Infection Can Cause Disease

In 1884 F. A. J. Löffler, using Koch's postulates, proved that the disease diphtheria was caused by a bacterium, which he named *Corynebacterium diphtheriae.* He showed that pure colonies of these bacteria introduced into test animals produced a fatal infection, with respiratory failure due in part to a membrane that forms across the respiratory tract. The animals had extensive damage of the heart, liver, kidney, and other organs, but bacteria could be cultured only from local lesions in the respiratory tract. Löffler postulated that a soluble toxin produced by the bacteria caused the widespread tissue damage.

By 1888 E. Roux had isolated this toxin, and by 1890 E. von Behring and S. Kitasato had succeeded in immunizing animals with an inactivated toxin, preventing the disease. The first child with diphtheria to receive the antitoxin was inoculated on Christmas night in 1891 in Berlin. By the 1920s an excellent and effective toxoid was made and widely used, eliminating this terrible bacterial disease. It was one of those great success stories where all the answers were in and we had won.

It came as quite a surprise, then, when in 1951 V. J. Freeman discovered that all the toxin-producing strains of *Cornyebacterium diphtheriae* were lysogenic and infected with a bacteriophage called beta. If the bacterial strains lost the beta phages, they no longer produced the toxin and were not virulent. Like all viruses, the beta phages are obligate intracellular parasites. Both bacteria and phages, then, are necessary to produce disease. In fact, the gene that encodes the information for the diphtheria toxin is a viral gene not present in the bacterial chromosome. The bacteria play a role in the expression of the phage toxin gene, but a viral gene causes the disease and, correspondingly, yields the antitoxin. In this example, d'Hérelle's fondest hope—that phages could be used to cure bacterial diseases—is realized, but not quite in the manner he expected. All things seem possible in the world of the viruses and their hosts.

The Evolution and Spread of a Virus

The yellow fever virus played a central role in the history of virology. Walter Reed and the U.S. Army Commission, in 1900, identified it as the first human virus to be isolated. Reed's team went on to prove that the virus was transmitted by mosquitoes and that eliminating the breeding grounds for these insects eliminated the disease.

Yellow fever is a zoonotic disease, which means that it has a major animal reservoir. In central Africa, where it is thought to have originated, wild nonhuman primates such as howler, owl, spider, and squirrel monkeys are infected. The virus

Major Walter Reed.

aegypti. This mosquito cannot breed in natural water holes or tree holes; it much prefers barrels, cisterns, shallow basins, and artificial containers. Yellow fever virus, however, adapts to the new mosquito host, which now becomes a vector of disease in human populations, spreading infection in villages and even urban areas.

In this way, the virus escaped from Africa in the seventeenth, eighteenth, and nineteenth centuries—principally on slave-trading ships. The virus came aboard ship in infected persons or in *Aedes aegypti*, which could breed in the water casks on the vessels. During long voyages, each round of breeding mosquitoes introduced new episodes of this epidemic; each time, with no land in sight, the source of this disease on shipboard appeared more and more mysterious.

From the seventeenth to the early twentieth century, major ports on the East Coast of the United States as far north as Philadelphia, New York City, and Boston were hit by yellow fever epi-

replicates in many organs of their bodies and spends time in the bloodstream. The aedine mosquitoes, like *Aedes africanus*, breed and lay their eggs in tree holes; they feed in the forest canopy, taking blood meals from the monkeys to provide nutrients for their developing eggs. The virus is taken up with the monkey blood and replicates in the mosquito, principally in the cells of the gut. When the mosquito bites again, the virus is in the saliva, which the insect regurgitates into the wound to prevent coagulation. Thus the virus circulates from a primate host to an insect and replicates in both, even though they are very different types of animals.

This cycle, termed jungle yellow fever, is disrupted when humans enter the jungle to clear or farm and later leave for urban environments. Humans infected by mosquito bites may have quite a severe disease, with 20 to 40 percent mortality; but when infected people return to towns and cities, they encounter a new species of mosquito, *Aedes*

Squirrel monkey in the tropical forest canopy.

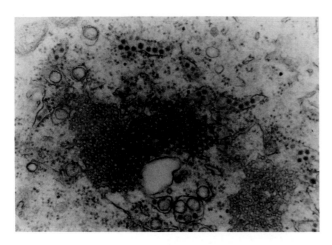

EM of a cell infected with Saint Louis encephalitis virus, a flavivirus closely related to yellow fever virus. Dense aggregates of ribonucleoprotein virus particles accumulate in infected cells; some capsids are seen here in vesicles, preparing to move to the cell membrane for maturation.

demics brought from Africa on ships transporting people, insects, and viruses. The yellow fever virus is now commonly found throughout South and Central America, where it did not exist previously. The movement of populations, the ability of the virus to move into a new species of mosquito with very different habits, and our poor understanding of the mechanisms and consequences of our behavior all contributed to the spread of a virus into virgin populations.

An Experiment in Virus Evolution

The myxoma and fibroma viruses—which cause tumors of gelatinous and fibrous connective tissue, respectively—belong to the poxvirus group, closely related to smallpox virus. In their natural hosts in North and South America, these viruses produce benign warts or skin tumors in rabbits, hares, and squirrels, circulating largely as unobserved, fairly benign disease agents in the wild populations. Exposure of a domestic rabbit or a wild European rab-

bit (both are different species from the New World wild rabbit) to the myxoma virus results, however, in severe disease symptoms, causing death from myxomatosis in 90 to 99 percent of the cases. This is another example of a virus that has come to an equilibrium, living with its natural host without doing too much damage. When it enters a new population, on the other hand, it can be lethal.

The wild European rabbit was first introduced to Australia in 1859. It very rapidly spread over the southern half of the continent, where it became a major pest in agricultural and grazing areas. The situation got so bad that, when other methods to keep the rabbit population under control failed, introduction of a lethal myxoma virus from the Americas was tried in 1950. The virus was shown to be restricted to rabbit populations—that was critical—and was spread by a mosquito biting the host. The original virus strain killed more than 99 percent of the infected animals, and the first few years after its introduction saw an enormous decline in the rabbit population.

The virus spread efficiently during the spring and summer, when mosquitoes were abundant, but the incidence and spread of disease were poor in each cold season because of the paucity of insects. In some places the virus even died out over the winter because of the lack of infected rabbits and poor transmission; but on a continentwide basis, the disease and the virus persisted. During each winter, rabbits that were infected with the most virulent virus died, so this most lethal strain of myxoma virus was not efficiently delivered to mosquitoes the next spring.

By contrast, some mutations in the virus created less virulent strains, permitting its hosts longer life and a better chance to survive the winter: these rabbits were available in the spring for mosquitoes to bite, thereby transmitting the less lethal disease. The requirement for survival over the winter months imposed a strong selection for a less lethal virus. Attenuated strains appeared in the spring of the very first year after the introduction of the myxoma virus into Australia, and three or four years later they were dominant.

European wild rabbits around an Australian water hole at dusk.

Rabbit populations infected with this less virulent virus began to show herd immunity—a phenomenon in which infection of a rabbit already immunized by previous exposure to a virus neutralizes that virus and lowers its probability of transmission to other animals, even if they are not immune. This accelerated the loss of the virulent strain, and the rabbit population resurged.

The rabbits that now bred were veterans of the initial exposure to the highly virulent myxoma virus. Among this group were rabbits that had survived because they were genetically resistant to the virus. The reasons for genetic resistance are complex (an unknown number of genes in specific combinations are thought to be involved), but such rabbits appeared quite rapidly. Within seven years after the introduction of the most virulent myxoma strain, which killed 90 to 99 percent of rabbits in the field, the same virus reintroduced into a population of rabbits that was not immune—a group with no previous exposure—killed only 25 percent. The difference was due to genetic factors in the host rabbit that had been selected for and were present in most individuals seven years after selection began. This experiment demonstrates that the most successful virus is one that can replicate many times in its host but is not recognized and causes little or no damage. These requirements are difficult to achieve, however, and most viruses do not fall into this category.

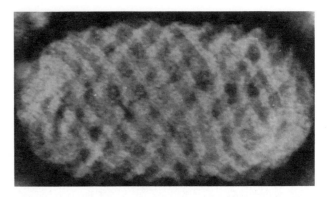

EM showing the brick shape and surface structure of the Orf virus, which is in the same subfamily of poxviruses as the myxoma viruses.

The Origins of Some Viruses Remain a Mystery

Over the past thirty years, vaccine production and testing have been carried out in cell culture, using tissue obtained from an organ—say, a kidney—placed in culture dishes to grow. The African green monkey from Uganda is commonly used as a source of such cells, which are prepared at various laboratories around the world. In 1967, twenty-five laboratory workers from three different locations—Marburg and Frankfurt in Germany and Belgrade, Yugoslavia—each processing monkey-kidney cell cultures, all contracted a similar disease: hemorrhagic fever. The patient typically has a very high fever, rash, and swelling followed by an uncontrollable bleeding in the organs, skin, and mucous membranes. As these patients were admitted to hospitals, six attending medical personnel contracted the disease, indicating human-to-human spread of an infectious agent. Among these thirty-one cases, there were seven deaths.

A virus was isolated from the blood and tissues of these patients, and extensive tests showed that it was unrelated to any known virus. The isolate caused a hemorrhagic-fever-like disease when inoculated into African green monkeys, and it was noted that several of the monkeys in a single shipment from Uganda had this hemorrhagic disease. It seemed likely that a monkey virus had crossed species, becoming more virulent, and attacked these human hosts. A large study of wild monkeys from the area of Uganda where these primates originated failed to detect any evidence of the isolate, now

called Marburg virus. The monkeys did not even have antibody against it, proving that these primates were not an animal reservoir for this virus.

Not a single additional case of hemorrhagic fever caused by Marburg virus was reported worldwide for eight years. Then, in 1975, three cases occurred in Johannesburg, South Africa. The first patient, a young Australian, had been traveling in Zimbabwe shortly before developing hemorrhagic fever and dying. Seven days after he became ill, his traveling companion checked into the hospital with the same symptoms; seven days later, a nurse contracted the Marburg virus disease. Both survived. The route of travel of the first two patients was reconstructed, and animals in those areas were tested for possible virus contacts, but the source of infection remained obscure.

Five years later in western Kenya, a case of Marburg virus inducing hemorrhagic fever was reported; the patient died in Nairobi, and one of the attending doctors also became infected. Interestingly, the area where the patient appeared to contract the disease was near the Ugandan source where the monkeys from the 1967 outbreak originated. In 1982, a single case was reported in Zimbabwe, close to a place visited by the 1975 patients during their travels. In 1987 another isolated case

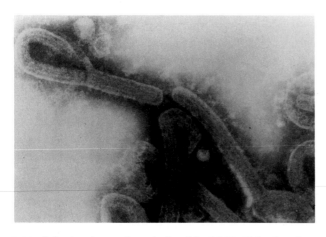

EM of the Marburg virus, isolated in 1967. This virus has appeared in five different years—1967, 1975, 1980, 1982, and 1987—causing miniepidemics with high rates of fatality and human-to-human infectivity.

occurred in western Kenya, near the location of the 1980 patient.

During this time, a closely related virus that also produced a hemorrhagic fever was detected. In 1976 in the Sudan and Zaire, an epidemic of about 550 cases occurred, with more than 430 deaths. This new virus was isolated and named Ebola virus, after a small river in Zaire. Electron micrographs showed it to be morphologically identical to Marburg virus, but antibodies made against it did not protect against Marburg virus: these isolates were related but distinct.

The epidemic spread of Ebola virus was by human contact, including sexual intercourse and common use of needles. After repeated human-to-human spread of the virus, the rates of fatality dropped. This suggested that, as the virus replicated in humans for a period of time (second, third, and fourth transmissions from person to person), it attenuated. This attenuation with passage (by mutation and selection) suggests an animal reservoir for the original case: a source outside the human population. In 1979 a second hemorrhagic fever epidemic occurred in the Sudan at the same site as in 1976. Ebola virus was recovered from these pa-

tients. Extensive tests of possible animal reservoirs have, to date, failed to find where the Ebola and Marburg viruses replicate and hide between epidemics. The true host-reservoir relationships are yet to be determined.

Conclusions

What have we learned from these stories and observations? First, we have surely been taught that there are rules that govern life processes; try as we may, we cannot violate them. All life forms are continually changing. Each generation brings new nucleotide sequences, information, and functions, and these are continuously tried out in various combinations in an ever-changing environment.

Some of these changes have been brought about by humanity within a remarkably short time frame, compared to the rates of biological change. Mankind has evolved into a unique species over the past one to four million years. Our viruses undoubtedly evolved along with us or adapted to us. Over the past ten to twenty thousand years, we have changed from a species living predominantly in small bands of food-gatherers and hunters to more urban settings where, in some cases, millions of us are packed together. The percentage of the space on our planet that we use has increased dramatically. Within four centuries, but especially in the past few decades, we have increased both the speed of travel and the number of individuals visiting new places. These cultural changes have surely had a profound impact upon our viral diseases. Clearly, we need to learn the consequences of our activities and decisions.

We have learned that all life processes follow the laws of chemistry and physics; molecules—be they replicating ones, informational ones, or nonreproducing chemicals—must abide by these rules. There is a difference, however, between events occurring in living organisms and those in the nonliving world. So long as an organism is capable of reproducing itself, rare events can be selected for and become the dominant form of life. When a rare event that cannot be replicated happens in the nonliving universe, it often remains minor in the field of observation. In contrast, a rare mutation that occurs in a virus only once in a million trials will, if it provides a replicative advantage, be selected for and become the virus of tomorrow. If we change the environment, we change the field against which new viruses are selected—in effect, we change the rules for selection, and new agents will certainly appear.

Finally, we have learned that viruses have multiple effects upon us. It has become clear that virus infections select the host that survives, just as we—by altering the field—select the virus that survives. We are what we are in part because we have survived the onslaught of our parasites. But the viruses we have studied have done more. They have contributed some of their nucleotide sequences to our own genetic endowment. We carry and pass to each generation the vestiges of retroviruses, integrated in our chromosomes and possibly exercising a sustained impact upon our selection and survival.

Viruses can be the conduit to move genetic information from one host to another. Sometimes this results in diseases as dramatic as cancer, as with the Rous sarcoma virus and its oncogene; and sometimes this may contribute to an organism's ability to survive, as seen in the T4 bacteriophage and its eukaryotic-like genes. That special relationship between host and parasite will continue to make human beings—and all forms of life on Earth—what we are and what we will be. It is important for us to know the rules.

Appendix

Classification of Viruses		
Name	Nucleic acid	Capsid symmetry
BACTERIOPHAGES		
T4	Double-strand DNA (same for T2 and lambda phages)	Head: elongated icosahedron Tail: helical, with spikes
T7	Double-strand DNA	Head: spherical icosahedron Tail: helical
ϕX174	Single-strand, circular DNA	Icosahedron
M13	Single-strand, circular DNA	Helical rod
Qβ	Single-strand, linear RNA	Helical rod
PLANT VIRUSES		
Tobacco mosaic virus	Single-strand (+), linear RNA	Helical protein rod
Tipular iridescent virus	Single-strand, linear RNA	Icosahedron
Viroids	Single-strand, circular RNA Small (240–375 nucleotides); base-paired	No protein coat
Virusoids	Single-strand, circular RNA Small (240–375 nucleotides); base-paired	No protein coat
ANIMAL VIRUSES		
HERPESVIRUSES	Double-strand (+/−), linear DNA 120,000–200,000 nucleotides	Icosahedron; protein coat; lipid envelope; 150–200 nm in diameter

Classification of Viruses

Name	Nucleic acid	Capsid symmetry
Virus	*Disease*	
Herpes simplex virus types 1 and 2 (HSV-1, HSV-2)	Recurrent cold sores, oral (type 1) or genital (type 2) lesions	
Varicella-zoster virus	Chicken pox and shingles	
Epstein-Barr virus (EBV)	Infectious mononucleosis; also associated with selected cancers in China and Africa	
Cytomegalovirus (CMV)	Several birth defects and, in special situations, pneumonia or hepatitis	
PAPOVAVIRUSES	Double-strand (+/−), circular DNA 5000−8000 nucleotides	Icosahedron; protein coat; 45−55 nm in diameter
Virus	*Disease*	
Human papilloma viruses	Types 16, 18, 31, 33, 35, and 39 associated with genital or oral carcinomas; types 6 and 11 associated with benign genital tumors	
Polyoma virus	Initiates tumors of many different types in the mouse	
Simian virus 40 (SV40)	A monkey virus that can initiate tumors in rodents	
ADENOVIRUSES	Double-strand (+/−), linear DNA 36,000−38,000 nucleotides	Icosahedron; protein coat; 70−90 nm in diameter
Virus	*Disease*	
Human adenoviruses	Some types can initiate tumors in rodents; in humans, respiratory or enteric disease, infectious pinkeye	
POXVIRUSES	Double-strand (+/−) DNA 130,000−280,000 nucleotides Virion includes RNA polymerase	Brick-shaped; 200−400-nm-long protein; lipids in the coat
Virus	*Disease*	
Smallpox virus	Smallpox	
Myxoma, fibroma viruses	Benign tumors, warts	
HEPADNAVIRUSES	Part single-strand, part double-strand (+/−), circular DNA 3300−3400 nucleotides Virion includes DNA polymerase and reverse transcriptase	Nucleocapsid; protein coat; lipid envelope

Classification of Viruses

Name	Nucleic acid	Capsid symmetry
Virus	*Disease*	
Hepatitis B virus	Serum hepatitis; chronic active carriers of the virus are at increased risk of liver cancer	
RETROVIRUSES	Two single-strand (+), linear RNA molecules per virion 3500–9000 nucleotides Virion includes reverse transcriptase (RNA to DNA)	Icosahedron; protein coat; lipid envelope; 80–130 nm in diameter
Virus	*Disease*	
Human T-cell leukemia virus-1 (HTLV-I)	Adult T-cell leukemia	
Human T-cell leukemia virus-2 (HTLV-II)	Possible link to hairy-cell leukemia	
Human immunodeficiency virus types 1 and 2 (HIV-1, HIV-2)	Acquired immunodeficiency syndrome (AIDS)	
A variety of animal retroviruses, including Rous sarcoma virus and avian leukosis virus	Associated with cancers in animals or immunodeficiencies in animals	
ORTHOMYXOVIRUSES	Eight single-strand (−), linear RNA molecules per virion 13,600 nucleotides total Virion includes transcriptase (−RNA to +RNA)	Helical; nucleocapsid; lipid envelope; 90–120 nm in diameter
Virus	*Disease*	
Influenza A virus	Respiratory disease	
PICORNAVIRUSES	Single-strand (+) RNA 7000 nucleotides	Icosahedron; protein coat; 28 nm in diameter
Virus	*Disease*	
Poliovirus	Infantile paralysis	
Rhinovirus	Common cold	
Hepatitis A virus	Infectious hepatitis	
TOGAVIRUSES	Single-strand (+) RNA 10,000–12,000 nucleotides	Icosahedron; protein coat; lipid envelope; 40–75 nm in diameter

Classification of Viruses

Name	Nucleic acid	Capsid symmetry
Virus	*Disease*	
Yellow fever virus	Yellow fever	
Hepatitis C virus	Hepatitis non-A non-B	
RHABDOVIRUSES	Single-strand (−) RNA 12,000 nucleotides Virion includes transcriptase (−RNA to +RNA)	Bullet-shaped; protein coat; lipid envelope; 60–175 nm long
Virus	*Disease*	
Rabies virus	Rabies	
PARAMYXOVIRUSES	Single-strand (−) RNA 15,900 nucleotides Virion includes transcriptase (−RNA to +RNA)	Helical; protein coat; lipid envelope; 125–250 nm in diameter
Virus	*Disease*	
Mumps virus	Mumps	
Measles virus	Measles	
REOVIRUSES	Double-strand (+/−) RNA Ten chromosomes; 1000–4000 base pairs	Icosahedron; protein coat; 70 nm in diameter
Virus	*Disease*	
Rotaviruses	Infant enteritis	
PARVOVIRUSES	Single-strand (+ or −) DNA 1000–2000 base pairs	Icosahedron; protein coat
Virus	*Disease*	
Adeno-associated viruses	Human, no known disease	

Sources of Illustrations

All line drawings are by Hans & Cassidy, Inc., except as noted below.

Chapter 1 *Facing p. 1:* A detail from *Vase of Flowers,* Jan van Huysum, 1772; The J. Paul Getty Museum. *p. 1:* Dr. Gopal Murti, Science Photo Library; Photo Researchers, Inc. *p. 2:* Dr. Hans Gelderblom, Robert Koch Institute, Berlin. *p. 3:* Rijkmuseum-Stichting, Amsterdam. *p. 4:* (bottom) Reunion des Musées Nationaux. *p. 5:* (top) Musée Pasteur, Institute Pasteur. (bottom) Deutsches Museum, München. *p. 6:* (top) Manfred Kage/Peter Arnold, Inc. *p. 7:* Dr. W. H. R. Langridge and Dr. A. A. Szalay, Plant Molecular Genetics and Biotechnology Center, University of Alberta. *p. 8:* John T. Finch, Medical Research Council Laboratory of Molecular Biology, Cambridge, England. *p. 9:* Dr. Michel Wurtz, University of Basel; (top) micrograph by J. Meyer. *pp. 10,11:* Dr. Babu Venkataraghavan, Lederle Laboratory, Pearl River, New York. *p. 12:* From Robert Snyder, *Buckminster Fuller, An Autobiographical Monologue/Scenario,* St. Martin's Press. *pp. 14,15:* From J. Darnell, H. Lodish, and D. Baltimore, *Molecular Cell Biology,* 2d ed., W. H. Freeman, 1991. *p. 17:* Dr. Carl-Henrik von Bonsdorff, Department of Virology, University of Helsinki. *p. 18:* Dr. Michel Wurtz, University of Basel. *p. 19:* Dr. Frederick A. Murphy, Centers for Disease Control (CDC), Atlanta, Georgia. *p. 20:* Dr. Michel Wurtz, University of Basel; micrograph by J. v.d. Broek. *p. 21:* Dr. Michel Wurtz, University of Basel; micrograph by M. Maeder. *p. 22:* (top) Dr. S. Rozenblatt and Dr. C. Moore, Tel Aviv University. (bottom) Dr. Carl-Henrik von Bonsdorff, Department of Virology, University of Helsinki.

Chapter 2 *p. 24:* Musée Pasteur, Institute Pasteur. *p. 25:* Dr. J. R. Paulson and Dr. M. L. Wong, University of California at San Francisco Medical School. *p. 26:* From Gunther S. Stent, *Molecular Biology of Bacterial Viruses,* W. H. Freeman, 1963. *p. 27:* From Darnell, Lodish, and Baltimore, *Molecular Cell Biology. p. 28:* Dr. Michel Wurtz, University of Basel; (B) micrograph by B. ten Heggeler. *p. 30:* (top) A. Marmont and E. Damasio, Division of Hematology, St. Martino's Hospital, Genoa. (bottom) From Darnell, Lodish, and Baltimore, *Molecular Cell Biology. p. 31:* California Institute of Technology Archives. *p. 33:* Cold Spring Harbor Laboratory Archives. *p. 34:* (bottom) Dr. Michel Wurtz, University of Basel. *pp. 35, 36, 37:* Dr. Babu Venkataraghavan, Lederle Laboratory. *pp. 38, 43:* Dr. Michel Wurtz, University of Basel. *p. 45:* Dr. Babu Venkataraghavan, Lederle Laboratory.

Chapter 3 *p. 46:* Freer Gallery of Art, Smithsonian Institution; translation by A. P. Hamori, Princeton University. *p. 47:* Dr. Frederick A. Murphy, CDC, Atlanta. *p. 48:* From Darnell, Lodish, and Baltimore, *Molecular Cell Biology. p. 49:* Ernst Haas/Magnum. *p. 53:* Dr. Babu Venkataraghavan, Lederle Laboratory. *p. 54:* Dr. W. Chiu and B. V. V. Prasad, Baylor College of Medicine, Houston. *p. 55:* After Darnell, Lodish, and Baltimore, *Molecular Cell Biology. p. 56:* Dr. Frederick A. Murphy, CDC, Atlanta. *p. 57:* National Library of Medicine. *pp. 58, 59:* Wellcome Institute for the History of Medicine. *p. 60:* World Health Organization. *p. 63:* March of Dimes Birth Defects Foundation. *p. 64:* Ny Carlsberg Glytothek, Copenhagen.

Chapter 4 *p. 66:* CDC/Science Source/Photo Researchers. *p. 67:* Dr. Bernard Roizman, University of Chicago. *pp. 69, 70:* Dr. Bernard Roizman, University of Chicago; from *Field's Virology,* Raven Press, 1990. *p. 71:* (left, A; right) Dr. Frazer J. Rixon, MRC Virology Unit, Institute of Virology, Glasgow. (left, B) Dr. Bernard Roizman, University of Chicago; *pp. 72, 73:* Dr. Bernard Roizman, University of Chicago; from *Field's Virology. p. 74:* Dr. Frazer J. Rixon, Institute of Virology, Glasgow. *p. 75:* Dr. Babu Venkataraghavan, Lederle Laboratory. *p. 77:* The Bettmann Archive. *p. 78:* After D. Burkitt, *Nature* 194 (1962). *p. 79:* The Children's Hospital of Philadelphia; photograph by Fritz Henle. *p. 81:* Dr. J. Martinez, Department of Molecular Biology, Princeton University. *p. 83:* Dr. Frederick A. Murphy, CDC, Atlanta.

Chapter 5 *p. 86:* Field Museum of Natural History. *p. 87:* Dr. J. T. Finch and Dr. A. Klug, MRC Laboratory of Molecular Biology, Cambridge, England. *p. 89:* Rockefeller Archive Center. *p. 92:* Dr. Baker and Dr. N. H. Olsen, Purdue University, West Lafayette, Indiana. *pp. 99, 100, 101:* Dr. Gerard

Zambetti, Department of Molecular Biology, Princeton University. *p. 104:* (A) Dr. Frederick A. Murphy, CDC, Atlanta. (B) Dr. Richard Feldmann, Division of Computer Research and Technology, National Institutes of Health (NIH), Bethesda, Maryland. *p. 108:* Fox Chase Cancer Center; photograph by Paul Cohen.

Chapter 6 *p. 112:* Joe Viesti/Viesti & Assoc. *pp. 113, 114:* G. H. Smith, National Cancer Institute, NIH. *p. 117:* Dr. Hans Gelderblom, Robert Koch Institute, Berlin. *p. 119:* Dr. Frederick A. Murphy, CDC, Atlanta. *p. 125:* Moravian Museum, Brno.

Chapter 7 *p. 130:* Marcel Miranda III/The Names Project. *p. 131:* Dr. Hans Gelderblom, Robert Koch Institute, Berlin. *p. 137:* National Cancer Institute, NIH. *p. 138:* G. W. Willis, Biological Photo Service. *p. 140:* (top) Dr. Thomas Folks, CDC, Atlanta. (bottom) Dr. Hans Gelderblom, Robert Koch Institute, Berlin. *p. 142:* (A) Dr. David Hockley, National Institute of Biological Standards and Control, South Mimms, England. (B) Dr. Hans Gelderblom, Robert Koch Institute, Berlin. *p. 143:* Nigel Dennis/Natural History Photo Agency. *p. 145:* Dr. Hans Gelderblom, Robert Koch Institute, Berlin. *pp. 146, 147, 148:* Dr. Carl O'Hare, New England Deaconess Hospital, Boston. *pp. 149, 150:* Dr. Babu Venkataraghavan, Lederle Laboratory. *p. 152:* From "Visual AIDS: An International Exhibition of AIDS Posters," curated by Prof. James Miller, University of Western Ontario, sponsored by London Life; photograph by John Tamblyn.

Chapter 8 *p. 154:* From Alesandra Comini, *Egon Schiele's Portraits,* University of California Press, Berkeley. *p. 155:* Peter Palese, Mount Sinai Medical School, New York City. *p. 156:* National Archives. *pp. 157, 160:* Charles D. Humphrey, Ultrastructure Group, DVRD, CDC, Atlanta. *p. 164:* Dr. Babu Venkataraghavan, Lederle Laboratory. *p. 169:* National Archives. *p. 173:* Osterreichesche Gallerie, Vienna. *p. 174:* The Bettmann Archive.

Chapter 9 *p. 176:* Jim Brandenburg/Minden Pictures. *p. 177:* Charles D. Humphrey, Ultrastructure Group, DVRD, CDC, Atlanta. *p. 178:* Dr. Michel Wurtz, University of Basel. *p. 182:* National Archives. *pp. 183, 187:* Charles D. Humphrey, Ultrastructure Group, DVRD, CDC, Altanta. *p. 192:* Merck & Co., Inc.

Chapter 10 *p. 194:* E. T. Archives. *p. 195:* Dr. Frederick A. Murphy, CDC, Atlanta. *pp. 200, 201:* From T. O. Diener, *Scientific American* 244 (1981). *p. 202:* W. E. Schadel, Small World Enterprises/Biological Photo Service. *p. 203:* From Diener, *Scientific American* 244 (1981). *p. 206:* (left) Dr. M. L. Wong and Dr. J. Paulson, University of California at San Francisco Medical School. (right) R. C. Williams, University of California at Berkeley, and B. Linquist, University of Olso. *p. 208:* (top) National Library of Medicine. (bottom) Stephen Dalton/Natural History Photo Agency. *p. 209:* Dr. Frederick A. Murphy, CDC, Atlanta. *p. 210:* Australian Overseas Information Service. *pp. 211, 212:* Dr. Frederick A. Murphy, CDC, Atlanta.

Index

monkey cells, use to grow poliovirus for vaccine, 91

monkey-kidney cell cultures, as source of hemorrhagic disease, 211

monkeys
AIDS-like disease in, 115
influenza A virus of, 168
as possible source of HIV viruses, 143–144
retroviruses of, 114, 115
simian AIDS (SAIDS) viruses of, 115
simian virus 40 of, 91–102
yellow fever in, 207

monocytes
CD4 protein on, 144
HIV infection of, 145

Montagnier, Luc, 139

Montague, Lady Mary Wortley, 58

mosquitoes
as myxoma virus vectors, 209
as virus vectors, 77, 83, 207–209

Mo T-cell line, HTLV-II isolation from, 136

mouse cells, SV40 culture in, 100

mouse mammary tumor virus, as retrovirus, 115

mouth lesions, from herpesvirus, 67, 73, 216

M-phase, of cell cycle, 126, 128

M1 protein, of influenza A virus, 158

M2 protein, of influenza A virus, 158, 160

mRNA
of adenoviruses, 106
of bacteriophage lambda, 43–45

copying of viral DNA by, 40, 44
factors affecting regulation of, 126–127
of hepatitis B virus, 184, 185, 187
of HIV, 145
of HSV, 69, 70, 72
of HTLV-I, 135
of influenza A virus, 160
of retroviruses, 113
role in transcription, 37, 38
role in virus origin, 196, 203
splicing by, 196
of SV40, 94, 96, 97
T-antigen transcription by, 94, 96
tumor virus DNA transcription into, 88–89
from virus-induced cancer cells, 91

MS-1, as early name for hepatitis A virus, 182

MS-2, as early name for hepatitis B virus, 182

mucosal candidiasis, in AIDS patients, 138

mumps virus, 218
vaccine for, 61, 62

Murakami, S., 183

murine sarcoma virus, ras oncogene of, 128

mutagens, effect on lysogenic bacteria, 45

mutations
in bacteriophage genes, 41
benefits of, 171
cancers induced by, p53 and Rb as targets in, 111
caused by reverse transcriptase, 147–148
as cause of immunocompromisation, 56
during cell division, 190

and chronic hepatitis B virus infections, 190
in defective retroviruses, 114
high rate of, in RNA, 172
in host cells, effect on virus attachment, 20
in HSV-1, 71–72
in human cancer, 129
of influenza A virus, 157, 166, 169–172
of oncogenes, human cancer from, 84
in p53 gene, 110, 129
in poliovirus, 64
in retinoblastoma gene, 108–109
role in B-cell lymphomas, 122
in SV40 enhancer nucleotides, 94
ultraviolet light induced, 102
of viruses, 20, 47, 48, 196, 204, 213

myc gene, 124, 128
deregulation of, in cancer induction, 123, 124
expression of, in cancer, 84, 122
as target of chromosome translocations, 122
as transcription factor, 128

myxoma viruses, 216
evolution of, 209–210

NAMES Project AIDS memorial, 131

nasopharyngeal carcinoma
Epstein-Barr virus association with, 82–83, 87
in southern China, 82–83
in United States, 83

National Institute for Medical Research (London), 162

National Institutes of Health, 132, 140–141, 182

nef gene, of HIV, 145, 147

negative regulators of growth, certain genes as, 125

negative sequence, of DNA nucleotides, 38

neuraminidase
of influenza A virus, 158, 159, 160, 163
subtypes of, 163

neurological disorders, caused by retroviruses, 115, 120

neurons, latent HSV-1 and HSV-2 in, 68–69, 157

neutralizing antibodies
directed against HA protein, 162
HBsAg induction of, 191

newborns
AIDS in, 141, 142
hepatitis B virus transmission to, 181, 189, 190, 192, 193
herpes infections in, 67

New Guinea, Burkitt's lymphoma incidence in, 76, 82

New Jersey Department of Health, 173

New Orleans, HTLV-II incidence in, 137

New York City
AIDS incidence in, 138
yellow fever epidemic in, 208

New Zealand, 190

Nirenberg, M., 37

North America
hepatitis B virus in, 190
myxoma and fibroma viruses in, 209

Rous-associated viruses, as retroviruses, 115
Rous sarcoma virus
  additional retroviral genes in, 117
  discovery of, 78
  as retrovirus, 114, 115, 217
  in short-latency cancer, 123
  src gene of, 123–128
  temperature-sensitive mutations of, 123
Roux, E., 207
Rozenbaum, Willy, 139
rubella virus, vaccine for, 61, 84
Russia
  hepatitis B virus in, 190
  influenza A virus in, 156
Rwanda, AIDS in, 142
Ryukyuans (Japanese ethnic group), HTLV-I in, 133–134

Sabin, Albert, 174
Sabin vaccine, 61, 62, 64
  Salk vaccine compared to, 65
sacral ganglia, latent HSV-2 in, 69, 75
Saint Louis encephalitis virus, 209
saliva, hepatitis B virus transmission by, 180, 181, 182, 188, 190
Salk, Jonas, 47, 63
Salk vaccine, 61, 63, 64
  Sabin vaccine compared to, 65
  SV40 as accidental contaminant in, 91–92
Salpêtrière Hospital, 139, 140
salvage pathway for nucleotide synthesis, 39

San Francisco, AIDS incidence in, 138
sanitation standards, role in poliovirus epidemics, 63–64
S antigen, of hepatitis B virus. See HBsAg
sarcoma virus, discovery by Peyton Rous, 78
Scandinavia, hepatitis B virus in, 190
schoolchildren, education for AIDS, 154
Schrödinger, Erwin, 32
seals, influenza A virus of, 162, 163, 168
secretory vesicle, of a cell, 15
semen, hepatitis B virus in, 188, 190
Semliki Forest virus, 16, 17, 22
sense sequence, of DNA nucleotides, 38, 94
serum hepatitis, from hepatitis B virus, 180, 181, 217
sexual transmission
  of AIDS, 139, 141, 155–156
  of hepatitis B virus, 181, 190
  of HPV, 102
  of HSV-2, 68, 69, 155–156
  of HTLV-I, 133, 134, 155–156
S gene, of hepatitis B virus, proteins encoded by, 184–185, 187
Shakespeare, 67
Shiga dysentery bacillus, 26, 28, 29
shingles, from varicella-zoster virus, 75, 76, 216
Shope, Richard, 88, 89, 162

Shope papilloma virus, 88
sialic acid, HA protein binding to, 158, 159, 162
sickle-cell anemia, hemoglobin defect in, 30–31
signal transduction pathway, in cell division regulation, 98, 129
simian AIDS (SAIDS) viruses, as retroviruses, 115
simian immunodeficiency virus (SIV)
  HIV-2 similarity to, 143
  as retrovirus, 115
simian sarcoma virus
  as retrovirus, 114
  sis oncogene of, 128
simian T-cell leukemia virus 1, 133
simian vacuolating virus 40. See SV40
simian virus 40. See SV40
Sindbis virus, 17
Sinsheimer, Robert L., 87
sis oncogene, 136
  of simian sarcoma virus, 128
skin carcinomas, 88
  HPV role in, 102
skin lesions, from herpesviruses, 67, 70
slave traders
  role in spread of smallpox, 57
  role in spread of yellow fever, 195, 208–209
slow viruses. See lentiviruses
small-plaque viruses, 48
smallpox, 75
  early efforts of protection against, 57–58
  eradication by vaccination program, 2, 56, 153
  history of, 57

from laboratory accident, 60
  role in Spanish conquest of New World, 2, 57, 156
smallpox virus, 216
  DNA sequences in, 57
  replication of, 57
  strains of, 57
  vaccine, 57–60, 181
Smith, J. R., 58
smooth endoplasmic reticulum, 16
Somalia, last case of smallpox in, 59, 60
sooty mangaby, retrovirus of, 143
South America
  hepatitis B virus in, 189
  myxoma and fibroma viruses in, 209
  yellow fever virus in, 209
Southeast Asia
  hepatitis B virus in, 189
  liver cancer in, 191
South Korea, lack of HTLV-I in, 134
Spanish flu epidemic, 165–166, 174
S-phase, of DNA synthesis, 125, 126
spinal cord trauma, shingles occurrence with, 76
"spontaneous generation," as early theory of microbe generation, 4, 203
src gene, 126, 135
  isolation of, 123, 125
  as oncogene, 128
Stahl, F., 36
Stanley, Wendell, 8
staphylococcal bacteria, 29
Stehelin, D., 123
sterile techniques, in medicine, 5

Other books in the Scientific American Library Series